A Specialist Periodical Report

# Environmental Chemistry

Volume 1

A Review of the Recent Literature Concerning the Organic Chemistry of Environments Published up to mid-1973

Senior Reporter:
**G. Eglinton,** *Organic Geochemistry Unit, School of Chemistry, University of Bristol*

Reporters
**J. E. Allebone,** *Department of Chemistry, Liverpool Polytechnic*
**P. A. Cranwell,** *Freshwater Biological Association, Ambleside, Westmorland*
**F. Culkin,** *Institute of Oceanographic Sciences, Godalming, Surrey*
**J. W. Farrington,** *Chemistry Department, Woods Hole Oceanographic Institution, Woods Hole, Mass, U.S.A.*
**P. Given,** *College of Earth and Mineral Sciences, Pennsylvania State University University Park, Penna., U.S.A.*
**R. J. Hamilton,** *Department of Chemistry, Liverpool Polytechnic*
**P. A. Meyers,** *Department of Atmospheric and Oceanic Science, University of Michigan, Ann Arbor, Michigan, U.S.A.*
**R. J. Morris,** *Institute of Oceanographic Sciences, Godalming, Surrey*
**B. Ravenscroft,** *Department of Chemistry, Liverpool Polytechnic*
**M. M. Rhead,** *Department of Environmental Sciences, Plymouth Polytechnic*
**J. W. Smith,** *CSIRO, Division of Mineralogy, North Ryde, New South Wales, Australia*



The Chemical Society
Burlington House, London W1V 0BN

**ISBN**: 0 85186 755 3

*Printed in Northern Ireland at The Universities Press, Belfast*

# *Preface*

This is the first volume in a new biennial series of Specialist Periodical Reports on Environmental Chemistry. This first volume concentrates upon the organic aspects of the subject although in future volumes it is planned to include inorganic and other aspects of environmental chemistry. Volume 2 is scheduled to appear towards the end of 1976.

The current volume, being the first, naturally has to provide a good deal of background. It is more descriptive, less condensed and less rigid in format than most Specialist Reports. The period of literature coverage is the two or three years up to mid-1973, but in some chapters this extends to late 1973; however, much prior work is included to give an overview. There are many gaps in the current treatment which it is hoped to fill later. At the present time there is certainly no single, well-defined body of information or of research activity which might be termed "Environmental Chemistry" and this naturally leads to some difficulty in designing and producing highly-structured and inter-related reports. However, the term does conveniently encompass several fast-growing fields of research which merit serious consideration by chemists and other scientists. Very broadly, one may define Environmental Chemistry as the assessment of the distribution and interaction of elements and compounds in the environment, their modes of transport and their effect on biological and other systems. The natural chemistry and the pollution chemistry of environments are best treated together. Thus, the fluxes of natural and pollutant compounds in the environment are both subject to the same processes and laws. A unified approach strengthens both fields.

The authors have written for chemists and non-chemists involved in environmental studies. They have defined certain environmental terms which are in common use but may not be known to chemists new to the field. A few study areas, which are intriguing but short of chemical data have been included in the hope of stimulating the necessary research. The formulae of some relatively simple and well-known compounds have been included in order to assist specialists other than chemists.

The Report emphasises aquatic environments. Indeed, most types of aquatic environment have been discussed as they are important areas for environmental studies. They are complex ecosystems into which organic matter is contributed directly and indirectly by living organisms, geological sources and anthropogenic sources, such as industry. Sediments deposited within aquatic environments can be regarded as communal 'sinks' and, to some extent, 'banks' for natural products and for pollutants. Little is known of the fate of compounds which enter the sediments but micro-organisms, including bacteria, fungi, protozoa and algae must play a large part in effecting changes in the organic matter. They consume and degrade it and contribute their own biomass to the sediment. Chemical and microbiological factors are both important, and to some extent they are not separable, with one being dependent on the other. Environmental organic chemical studies have to inter-relate the organic carbon of the whole ecosystem. Studies need to cover a wide variety of environments, including those variously combining marine, freshwater, eutrophic, oligotrophic, arctic, sub-arctic, temperate, sub-tropical and tropical conditions. Current emphasis often lies on those environments under stress in the industrialised areas of the world.

Early work on the organic matter of aquatic environments was largely concerned with simple measurements, such as the total amount and distribution of organic matter as indicated by oxidation. The emphasis has now switched to molecular characterisation and quantification of individual compounds. The significance of this sort of work can be seen by examining the programmes involved in determining the distribution of hydrocarbons in the marine environment (Chapter 5). Quantitative data are being acquired rapidly but much is contradictory and difficult to integrate, primarily because of the difficulty in distinguishing between 'natural' hydrocarbons of biological and geological origin and 'pollutant' hydrocarbons contributed by man's activity in the form of crude oil spills and sewage. This area of research has direct relevance for marine environmental quality and off-shore drilling programmes and, inevitably, international politics. The arrangement of the chapters is as follows:

Chapter 1, 'Stable Isotope Studies and Biological Element Cycling,' by J. W. Smith, is concerned with the distribution of the stable isotopes of the light elements—carbon, sulphur, nitrogen, hydrogen, and oxygen—in environments. It surveys recent work on the biogeochemical cycling of these elements. Such studies are important guides to the operation of the natural cycles and to the effects of pollution. Environmental work in this area bridges organic and inorganic interests.

Chapters 2, 3 and 4 group together in that they are concerned with the chemistry of most of the major types of aquatic environment. Chemical classes are described in terms of their qualitative distribution patterns in the environment, their reaction pathways (mainly conjecture at this point) and, to some extent, their overall budgets. Analytical techniques are included here since they are essential to an understanding of the type of data being obtained. Each chapter contains some discussion of appropriate aspects of biochemistry, natural product chemistry, chemical ecology, and microbiology. There are also points of contact with the organic geochemistry of ancient sediments, including crude oil and coal, mineralogy and petrology, and colloid science. The involvement extends to industrial chemistry, because of the products released into the environment, and to the physics and chemistry of transportation processes. For each environment, there is some discussion of the environment itself, its chemistry and of the kinetics involved in deriving a 'model' of its operation. Sediments are records of paleoenvironments and hence older sediments provide reference points for current environmental conditions.

The topics covered by the above three chapters are: Chapter 2, 'Rivers and Lakes, Both Water and Sediment' by P. A. Cranwell; Chapter 3, 'Bogs, Marshes, and Swamps' by P. H. Given; and Chapter 4, 'Oceans, Fjords, and Anoxic Basins' by R. J. Morris and F. Culkin. Cranwell's review of the freshwater bodies has relevance for water resource management and environmental conservation. Control of water quality needs information on the input of toxic or unpleasant substances, either by pollution or by natural processes such as the growth of algal blooms. Given's treatment of wetland environments is also relevant for water supplies and conservation. It deals with productivity in food chains and has especial relevance for metal–organic interactions and the origin of coal and peat. The effect of human activity on wetlands is illustrated by the changes which have taken place in the Everglades from 1871 to 1971. Given also points out the significance of the wetlands as a site for the escape of organic matter from the carbon cycle, through accumulation in the water-logged environment. Morris and Culkin's treatment of the oceans *etc.* reveals that interesting distributions of chemical compounds are observed and that the really important boundaries are the air/water and water/sediment interfaces.

A different treatment is used in Chapters 5, 6, and 7. Here, we have taken a particular, environmentally-important class of compound and examined the methods for its analysis and the determination of its distribution and fate in environments. This in-depth treatment cuts across environmental boundaries and complements that of surveying all types of compound in a single environment. Thus, in Chapter 5, 'Hydrocarbons in the Marine Environment' Farrington and Meyers point out that research is proceeding at a very fast rate, interest being generated by the effects of oil pollution. There is a major contamination problem in studying hydrocarbons, which is bound up with biosynthesis and natural product chemistry, geochemical processes, and anthropogenic effects such as urban and industrial pollution. Chapter 6, 'The Fate of DDT and PCB in the Marine Environment' by M. M. Rhead, takes another very well-known group of compounds, the chlorinated hydrocarbons, and examines their fate in the same environment. This is now a classic environmental topic but a full understanding of the fate of these compounds will depend on an understanding of the fate of natural organic compounds. In Chapter 7, Allebone, Hamilton, and Ravenscroft examine the fate of one rather more readily degraded compound, 2,4-dichlorophenoxyacetic acid. The distribution and fate of this type of compound in the environment is here closely connected with its use in agriculture.

Future volumes will include reports on the chemistry of air pollution and of atmospheric processes involving carbon and other light elements. Major environments requiring treatment are soils, estuaries, and continental shelves. Similarly, there is some justification for treating public water supplies and sewage treatment plants as separate environmental problems. Small and very large molecules both deserve specific attention. Thus, element cycles involving carbon, hydrogen, nitrogen *etc.* have important links in the form of small molecules such as carbon monoxide, ethylene, acetylene, and ammonia. Pollutant studies should include low molecular-weight chloro- and fluoro-compounds. Fuller treatment of element cycles in terms of mathematical models is another important area for future reviews.

*July* 1974 G. EGLINTON

# Contents

# 1
# Stable Isotope Studies and Biological Element Cycling

BY J.W. SMITH

## 1 Introduction

Natural biological, physical, and chemical processes operating over geological time have resulted in the establishment of recognizable patterns in the distribution of the stable isotopes of many of the light elements. This knowledge and an increasing understanding in detail of the many individual processes involved in the creation of this pattern now allow the sources and previous histories of light elements in many geological systems to be determined with considerable certainty. Urey[1] first demonstrated the connection between the environment and isotopic ratios and developed the oxygen thermometer for the evaluation of palaeotemperatures. Since these early experiments the method has acquired increasing recognition and application. Very recently[2] the value of isotope-ratio measurements in revealing otherwise unobservable relationships and effects has been demonstrated in studies of the distribution of the light elements in returned lunar samples.

For the purpose of this discussion it must be assumed that the organic geochemist is primarily concerned with the isotopic composition of those organic compounds currently present, or being created or destroyed, in order that the biogeochemistry of natural processes may be better understood. However, much of the organic material in these three categories has recently been introduced into the present environment by man and it is therefore essential to know the extent and effect of such additions if a meaningful interpretation of experimental data is to be made. In this respect, the role of fossil fuels can rarely be ignored, a situation well demonstrated by the very considerable interest which continues to be paid to the effects on the environment of the direct release of either fossil fuels or the by-products resulting from their utilization in the chemical industry and power production. Even when due regard is paid to these effects, a meaningful understanding and interpretation of isotopic data can scarcely be made if interest is solely limited to organic molecules. Very often in Nature the immediate precursor of an organic compound is an inorganic molecule, an example being the photosynthesis of sugars from carbon dioxide, and, since the isotopic composition

[1] H. C. Urey, *J. Chem. Soc.*, 1947, Part 2, 562.
[2] I. R. Kaplan, *Space Life Sciences*, 1972, **3**, 383.

of the product is dependent on that of the reactant, it becomes essential in environmental studies to give some consideration to such inorganic portions of the element cycle. Perhaps the greatest benefit to be gained from isotopic measurements is the ability to determine both the precursors and decomposition products of materials of interest and as a result, biogeochemical studies commonly include not only investigations of the distribution and isotopic composition of existing organic compounds, but also of related inorganic species, *e.g.* sulphate, sulphide, and carbon dioxide, which may be of significance in the biological assimilation and cycling of the elements.

In a Report which is primarily concerned with organic materials, a full discussion of all those processes, both organic and inorganic, which result in a fractionation of the isotopes cannot be entertained. Accordingly, only those inorganic processes which most obviously and directly affect the distribution and isotopic composition of organic compounds are considered. It is understood, however, that all reactions which result in isotopic fractionation probably modify the isotopic ratios in organic compounds to some extent, even if this is not directly detectable. Not excluded are those conversions by microorganisms in which both the reactants and products are inorganic compounds and the organisms in fact provide little more than a pathway for the completion of thermodynamically favoured reactions. In the case of the dissimilatory bacterial reduction of sulphate, whilst at any stage the quantities of sulphur organically bound within cellular material are probably negligible when compared with the large quantities of sulphide produced, the major role played by this process in the sulphur cycle and the marked isotopic fractionations which result make the inclusion of such metabolic conversions essential.

## 2 Carbon

Since several excellent reviews of the geochemistry of the stable carbon isotopes are available,[3] it is sufficient that only brief mention be made here of the processes responsible for isotopic fractionation. Either directly or indirectly, biological materials result almost entirely from photosynthesis. Carbon in the forms of gaseous and dissolved $CO_2$ or as bicarbonate in solution may be utilized in the photosynthetic process; however, since at equilibrium the bicarbonate in solution is considerably enriched in $^{13}C$* relative to $CO_2$ in solution or in the gaseous state,[4] the isotopic composition of photosynthesized materials will vary with the source of carbon available. In Nature these two

* $\delta^{13}C‰ = \frac{[^{13}C/^{12}C]_{Sample} - [^{13}C/^{12}C]_{Standard}}{[^{13}C/^{12}C]_{Standard}} \times 1000$ where the standard is Peedee Belemnite.

[3] H. Craig, *Geochim. Cosmochim. Acta*, 1953, **3**, 53; E. T. Degens, in 'Organic Geochemistry', ed. G. Eglinton and M. J. Murphy, Springer-Verlag, Berlin, 1969; H. P. Schwarz, in 'Handbook of Geochemistry', ed. K. H. Wedepohl, Springer-Verlag, Berlin, 1969.

[4] W. G. Deuser and E. J. Degens, *Nature*, 1967, **215**, 1033; H. G. Thode, M. Shima, C. E. Rees, and K. V. Krishnamurty, *Canad. J. Chem.*, 1965, **43**, 582.

major sources of carbon are the atmosphere and bicarbonate in solution in the oceans and, in general, materials derived from these two reservoirs may be distinguished by their $^{13}C$ contents.[3,5] However, since the degree of isotopic fractionation between the two reservoirs decreases with temperature and the quantity of dissolved $CO_2$ relative to bicarbonate in solution decreases with pH, estimates of the environment during photosynthesis based on isotopic measurements are not always precise. Differences in isotopic composition also arise between the carbon source and the products during photosynthesis. This fractionation has been attributed to the relative collision rates of the $CO_2$ molecules with the leaf surface.[6] Detailed studies of the process[7] indicate that the major fractionation stage, which results in the photosynthetic product being enriched in $^{12}C$ by some 17‰ relative to atmospheric $CO_2$, commonly occurs during the enzymatic fixation of dissolved $CO_2$ as 3-phosphoglyceric acid.

Whilst the above situation holds in general for the majority of higher plants (that is, those which use the Calvin cycle in photosynthesis), evidence has been gathering to show the existence of other synthetic pathways for enzymatic fixation of carbon which give rise to different $^{13}C/^{12}C$ ratios in the final plant products.[8] A study of 104 selected species of plants[9] has revealed a much wider variation in the $^{13}C$ contents than might previously have been expected; many terrestrial mono- and di-cotyledons and one gymnosperm have $\delta^{13}C$ values greater than −18‰. Plants within this category included many from desert, salt-marsh, and tropical environments; where less favourable conditions for plant growth prevail it is suggested that the high $^{13}C$ contents in these plants may reflect the utilization of other more efficient photosynthetic cycles under these harsher conditions. Considerable variations in the $^{13}C/^{12}C$ ratios between sub-species growing in different environments are reported in support of the view that physiological adaptations to the environment have been made by the plants.

Variations in the $^{13}C$ contents of the products of photosynthesis also occur and commonly appear as isotopic differences between the extractable lipid portion of the plant and its main structure,[7,8,10] or within particular classes of chemical compounds, *e.g.* carbohydrates,[11] fatty acids,[12] and amino acids.

[5] F. E. Wickman, *Geochim. Cosmochim. Acta*, 1952, **2**, 243.
[6] H. Craig, *J. Geol.*, 1954, **62**, 115.
[7] R. Park and S. Epstein, *Geochim. Cosmochim. Acta*, 1960, **21**, 110; P. H. Abelson and T. C. Hoering, *Proc. Nat. Acad. Sci. U.S.A.*, 1961, **47**, 623.
[8] H. P. Kortschak, C. E. Hartt, and G. O. Burr, *Plant Physiol.*, 1965, **40**, 209; M. D. Hatch and C. R. Slack, *Ann. Rev. Plant Physiol.*, 1970, **21**, 141; B. N. Smith and S. Epstein, *Plant Physiol.*, 1970, **46**, 738; T. Whelan, W. M. Sackett, and C. R. Benedict. *ibid.*, 1973, **51**, 1051.
[9] B. N. Smith and S. Epstein, *Plant Physiol.*, 1971, **47**, 380.
[10] S. R. Silverman, in 'Isotopic and Cosmic Chemistry', ed. H. Craig, S. L. Miller, and G. T. Wasserburg, North-Holland, Amsterdam, 1964; J. A. Calder and P. L. Parker, *Geochim. Cosmochim. Acta*, 1973, 37, 133.
[11] E. J. Degens, M. Behrendt, B. Gotthardt, and E. Reppmann, *Deep Sea Res.*, 1968, **15**, 11.
[12] P. L. Parker, *Ann. Rep. Dir. Geophys. Lab. Carnegie Inst. Washington Year Book* 1961—2, **61**, 187.

Since the major sources of carbon for photosynthesis are of inorganic form, although they may have been immediately derived from organic materials, it is essential that reference be made in this review to those investigations in which efforts to relate organic and inorganic carbon are made Systems in which organic forms of carbon are not immediately involved wil not be discussed here.

In attempts to determine the origins of naturally occurring organic compounds, isotopic comparisons are frequently made with other organic compounds which have resulted from the biological utilization of either atmospheric carbon dioxide or those carbon forms that are in solution in sea water. In many instances such comparisons have proven to be rewarding, and consequently the continued interest in this approach results in fresh additions being frequently made to the already sizeable literature on this aspect of isotope chemistry. Thus, whilst it has long been recognized that humic acids in non-marine sediments result from the degradation of the lignin in land plants, only comparatively recently has it been shown that humic acids constitute a very considerable fraction of the organic matter in marine sediments.[13] Whether these marine acids are composed largely of transported continental materials, whether they are autochthonous and result from the recombination of the decomposition products of plankton, or whether they may be of dual origin is not fully resolved, although the general evidence favours the last view. Since terrigeneous plants are usually enriched in $^{12}C$ relative to marine plankton,[3,9] and it has been shown[3,14] that the isotopic composition of the organic matter in marine sediments varies from $\delta^{13}C$ $-19$ to $-22$‰ and largely reflects that of the plankton in the water, several investigators have measured the $^{13}C/^{12}C$ ratios of marine and non-marine organic residues in attempts to determine the sources of carbon in each and to differentiate between these.[15] Much of these data and those from their own studies of the humic acids from a wide range of marine, coastal, littoral, and continental sediments and soils has recently been combined by Nissenbaum and Kaplan[16] in an effort to resolve this problem finally. $\delta^{13}C$ values in the 20 marine samples examined range from $-17.2$ to $-27.4$‰, with these extreme values relating to materials from the Cariaco Trench and the Santa

[13] E. T. Degens, J. H. Reuter, and N. F. Shaw, *Geochim. Cosmochim. Acta*, 1964, **28**, 45; O. K. Bordovskiy, *Marine Geol.*, 1965, **3**, 33; V. I. Kasatochkin, O. K. Bordovskiy, N. K. Larina, and K. Cherkinskaya, *Doklady Akad. Nauk. S.S.S.R.*, 1968, **179**, 690.

[14] W. M. Sackett, *Marine Geol.*, 1964, **2**, 173; M. A. Rogers and C. B. Koons, *Trans. Gulf Coast Assoc. Geol. Soc.*, 1969, **19**, 529; R. S. Scalan and T. D. Morgan, *Internat. J. Mass Spectrometry Ion Phys.*, 1970, **4**, 267.

[15] V. E. Swanson and J. G. Palacas, Geological Survey Bulletin 1214-B, U.S. Government Printing Office, Washington D.C., 1965; J. G. Palacas, V. E. Swanson, and A. H. Love, Geological Survey Professional Paper 600-C, C97, U.S. Government Printing Office, Washington D.C., 1968; A. Otsuki and T. Hanya, *Geochim. Cosmochim. Acta*, 1967, **31**, 1505; A. Nissenbaum and I. R. Kaplan, *Chem. Geol.*, 1966, **1**, 295; M. A. Raschid and L. H. King, *Geochim. Cosmochim. Acta*, 1970, **34**, 193; F. S. Brown, M. J. Baedecker, A. Nissenbaum, and I. R. Kaplan, *ibid.*, 1972, **36**, 1185; W. M. Sackett, W. R. Eckelmann, M. L. Bender, and A. W. H. Be, *Science*, 1965, **148**, 235.

[16] A. Nissenbaum and I. R. Kaplan, *Limnology and Oceanography*, 1972, **17**, 570.

Monica Basin, respectively. The high $^{12}C$ content of the latter is explained by a large influx of land-plant material, but no explanation for the other anomalous extreme value is offered. When these two samples are excluded, an average value of −22.2‰ results, with a standard deviation of 1.0‰. The 12 coastal and littoral samples were found to have $\delta^{13}C$ values of from −19.1 to −27.3‰, with the 3 samples from tidal marshes being most enriched in $^{13}C$ and having values of −19.1, −19.3, and −21.2‰. The average value for the remaining 9 samples is −25.3‰ with a standard deviation of 1.0‰. The 14 continental samples exhibited the greatest variation in $^{13}C$ contents, with values of from −14.8 to −29.1‰ being reported. The highest $^{13}C$ content related to soil from a sugar-cane plantation in Hawai. Carbon fixation in cane is *via* the Hatch–Slack pathway, and inclusion of plant debris in the soil probably accounts for the high $^{13}C/^{12}C$ ratio.[8] No reason for the high $^{13}C$ content of a Hula peat sample is given (−19.2‰). The sharp isotopic difference between the sediment (−21.0‰) in land-locked Lake Haruna and the soil (−28.2‰) from the lake shores shows that the former originates from a lacustrine biota rather than land-plant materials.[15] When the three isotopically 'heavy' samples are excluded, the remainder have an average value of −26.0‰ and a standard deviation of 1.5‰.

Although a general, if not well-defined, differentiation between marine, coastal, and continental humic acids can be made on the basis of absolute isotopic composition, the significant number of samples which are not easily accommodated into these three classifications suggest that either the processes determining the isotopic composition of the samples examined are insufficiently understood, or additional processes are operating.

Isotopic measurements are also used to illustrate the fact that although a contribution of terrigeneous humic acids to marine deposits often occurs close to continental margins, in general these acids are seldom transported far into the oceans, except where high-energy turbidity currents are involved.

In contrast to this broad survey, the U.C.L.A.[17] group have recently reported their findings from a detailed 'in depth' study of the forms of carbon in samples of sediments and interstitial waters from several locations in Saanich Inlet, a fjord in British Columbia. The reported $\delta^{13}C$ values of −19.2‰ for the plankton, −26.6‰ for the humus-rich soil in the Inlet surroundings, and −20.1‰ to −22.5‰ for the marine sediments suggest a dual origin for the organic matter in the sediments, a view which is further confirmed by the distribution of lipid constituents in these. Measurements on various classes of extractable compounds in the sediments, soils, and plankton gave a consistent isotopic pattern (Table 1). In every case the products derived from the plankton were enriched in $^{13}C$ relative to the average values for the sediment and the products from the soils were depleted in $^{13}C$ content relative to the sediment, thus confirming the value of this approach in this case and the dual origin of the sedimentary material.

[17] A. Nissenbaum, M. J. Baedecker, and I. R. Kaplan, *Geochim. Cosmochim. Acta*, 1972, **36**, 709; A. Nissenbaum, B. J. Presley, and I. R. Kaplan, *ibid.*, p. 1007.

**Table 1** *Values of $\delta^{13}C$‰ for samples taken in the region of Saanich Inlet, British Columbia*

| | $\delta^{13}C$‰ | | |
|---|---|---|---|
| *Compound* | *Sediment* | *Soil* | *Plankton* |
| n-paraffins | −25.0 to −30.7 | −29.9 | −24.0 |
| Free fatty acids | −22.9 to −26.9 | −30.2 | −20.1 |
| Hydrolysable fatty acids | −23.6 to −26.4 | −29.8 | −24.5 |
| Hydrolysable amino-acids | −19.2 to −22.2 | −21.8 | −15.8 |
| Humic acids | −21.9 to −23.1 | −29.1 | — |

As much as 150 mg $l^{-1}$ of dissolved organic matter consisting of high-molecular-weight polymers of amino-acids and carbohydrates was extractable from the interstitial waters. The chemical and isotopic composition ($\delta^{13}C$ −20 to −21‰) of this material, which is believed to be the precursor of fulvic and humic acids, indicates that it results largely from the recombination of plankton degradation products, a conclusion which is in marked contrast to the widely held view that humic acids are derived from the lignin and cellulose derivatives of higher plants. Differences between the $^{13}C$ contents of these acids and the more highly condensed insoluble organic residues are thought to be largely due to the loss of isotopically 'heavy' $CO_2$ during decarboxylation reactions.

The distribution and isotopic composition of the other forms of carbon present in the Saanich Inlet samples are particularly interesting. $\delta^{13}C$ values for the sediment carbonates range from +1.0‰ at the surface to −3.5‰ at depth, a change which is attributed to the production of biogenic $CO_2$ in the deeper anoxic regions of the basin. However, $\delta^{13}C$ values for the dissolved $CO_2$ in the corresponding interstitial waters vary from −11‰ near the surface (one value of −37‰ is reported) to +18‰ at depth. If these high $^{13}C$ contents arose from a preferred utilization of the lighter isotope, both $CO_2$ and $^{12}C$ contents should decrease with depth, as in continental-shelf sediments.[18] Since this is not so, an explanation other than dependence on a simple kinetic effect is required.

The formation of isotopically 'heavy' $CO_2$ as the result of exchange between this and the methane present in the system is not an acceptable explanation, since this exchange is extremely slow relative to that between $CO_2$ and carbonate, and equilibration between the latter compounds was not established. It has been shown that $CH_4$ and $CO_2$, the latter strongly enriched in $^{13}C$, can be produced by the fermentation of acids[19] but, in view of the large quantities of $CO_2$ involved, the authors favour the reduction of preformed biogenic $CO_2$ ($\delta^{13}C$ −20‰) resulting from the diagenesis of the organic material present, by methane-forming bacteria using the molecular or organically available hydrogen in the system. Reduction of $CO_2$ by such methods

[18] B. J. Presley and I. R. Kaplan, *Geochim. Cosmochim. Acta*, 1968, **32**, 1037.
[19] W. D. Rosenfeld and S. R. Silverman, *Science*, 1959, **130**, 1658.

has been experimentally demonstrated[20] and the degree of isotopic fractionation is in agreement with kinetic data.

Similar measurements have been less helpful in determining the origin of the extractable organics in the Dead Sea.[21] The $^{13}C$ contents of the lake sediments (−23.8 to −24.3‰), surface plankton (−24.8‰), surrounding soil (−24.3‰), closely associated oil shale (−28.7‰), and asphalt (−26.0‰) indicate that the contribution of carbon from the two latter possible sources is insignificant, but still do not allow the origin to be determined.

Parker,[22] in an earlier study of shallow marine systems, has commented on the variations in isotopic composition which exist between organisms and between different compounds from the same organism using the same carbon source. The organic carbon in the individual organisms ranged in $^{13}C$ content, relative to the inorganic carbon in the seawater, from 0 to −20‰, and in every case the lipids or fatty acids were depleted in $^{13}C$ by from 4 to 15‰ relative to the total organic carbon in the organism. The author suggests that in view of these results caution must be exercised when attempts are made to relate biogenic residues to particular growth environments on isotopic evidence alone.[23] In the same system, diurnal variations of 4‰ in the $^{13}C$ content of the sea water were observed to correspond directly with the preferred utilization of $^{12}CO_2$ during photosynthesis by day and the respiration of $^{12}C$-enriched carbon dioxide throughout the hours of darkness. Similar changes have been described in the atmosphere over densly wooded areas and grasslands, where both $^{12}C$ and carbon dioxide contents fall during the day and rise at night.[24] Decreasing $^{13}C$ contents in city atmospheres as the result of vehicle exhaust-gas pollution have also been reported, as have been changes in the isotopic composition of wood samples with age as a result of increasing contributions of $CO_2$ from the combustion of fossil fuels.[25] CO production from combustion appears not to be of general significance. The photo-oxidation of methane is clearly the principal source of CO, although seasonal and local variations due to the autumnal death of plants, increased domestic heating, *etc.*, occur. Five sources of CO with $\delta^{13}C$ values from −22 to −30‰ are listed.[26]

The variability in the isotopic composition of the total inorganic carbon in estuaries and bays, as the result of changes in the contribution of freshwater carbon dioxide and pollution with petrochemicals and sewerage, has been contrasted with the constancy of the $^{13}C$ contents of the oceans and the atmosphere. Measurements on the dissolved organic carbon in the waters of the Houston Ship Channel indicate that almost 70% of this carbon is of

20 Y. Takai and T. Kamura, *Folia Microbiologica (Prague)*, 1966, **11**, 304.
21 A. Nissenbaum, M. J. Baedeker, and I. R. Kaplan, *Geochim. Cosmochim. Acta*, 1972, **36**, 769.
22 P. L. Parker, *Geochim. Cosmochim. Acta*, 1964, **28**, 1155.
23 P. L. Parker and J. A. Calder, *Inst. Marine Sci. University of Alaska*, 1970, **1**, 107.
24 C. Keeling, *Geochim. Cosmochim. Acta*, 1960, **24**, 299.
25 I. Friedman and A. P. Irsa, *Science*, 1967, **158**, 263; H. L. Dequasie and D. C. Grey, *Int. Lab.*, 1971, 20.
26 T. A. Maugh, *Science*, 1972, **177**, 338.

petrochemical origin and the value of such measurements, where the isotopic composition of the possible contaminant is known, is clearly demonstrated.[27] However, in other systems where the decrease in $^{13}C$ contents as the result of pollution with organic wastes is not so pronounced, difficulties in determining this isotopic change often result from a scarcity of data on the natural unpolluted values and natural variations in the isotopic composition of these.

Although isotopic fractionation occurs during skeletal carbonate formation by carbonate-secreting organisms, the bicarbonate reservoir available is usually unlimited and so no short-term environmental effect is observable. The shell carbonate of molluscs appears to form in isotopic equilibrium with the bicarbonate in sea water, and it isotopically resembles abiologically precipitated carbonate.[28] In contrast, the carbonate secreted by other organisms, *e.g.* corals and sea urchins, has variable $^{13}C$ contents.[29] The factors responsible for this apparently non-equilibrium process are still not clearly defined.

**Biological Cycling.**—The major, well-established, stages in the carbon cycle which facilitate transport and conversion of the various carbon forms and which result in isotopic fractionation include:

(*a*) The equilibration between atmospheric carbon dioxide, dissolved carbon dioxide and bicarbonate in the oceans, and precipitating carbonate. Discussions of the equilibrium distributions, and the pH and temperature dependence of these, have been outlined previously, the net result being an average depletion in atmospheric carbon dioxide of 7‰ relative to the $^{13}C$ content of marine carbonate.

(*b*) The preferred utilization of isotopically 'light' carbon dioxide during photosynthesis which results in the biogenic product being further depleted in $^{13}C$ by some 17‰ relative to the carbon dioxide source. Thus, when the atmosphere is virtually the only source of carbon, as is the case with land plants, these plants will generally be some 25‰ depleted in $^{13}C$ relative to marine carbonate, although variations due to the use of less common photosynthetic pathways occur. Isotopic differences, particularly between the lipid and non-lipid portions of the plant, produce further variations, and $\delta^{13}C$ values of $-23‰$ to $-28‰$ are customarily found for land plants. Correspondingly, since marine plants are able to utilize the isotopically 'heavier' bicarbonate in solution during photosynthesis, these plants are more enriched in $^{13}C$, and $\delta^{13}C$ values from $-18‰$ to $-22‰$ are common.

(*c*) Depositional processes involving biogenic residues. Where such residues are preserved, even only partially, diagenetic alteration of these will

[27] J. A. Calder and P. L. Parker, *Environ. Sci. and Technol.*, 1968, **2**, 535; P. L. Parker, in 'Impingement of Man on the Oceans', ed. D. W. Hood, Wiley, New York, 1971, p. 431.

[28] S. Epstein, R. Buchsbaum, H. Lowenstam, and H. C. Urey, *Geol. Soc. Amer. Bull*, 1951, **62**, 417.

[29] J. N. Weber and D. M. Rauf, *Geochim. Cosmochim. Acta*, 1966, **30**, 681; J. N. Weber and D. M. Rauf, *ibid.*, p. 704; J. N. Weber and P. J. M. Woodhead, *Chem. Geol.*, 1970, **6**, 93.

result in the formation of a carbonaceous shale or coal or crude oil, *etc.*, the end-product being dependent on the nature of the original material and the diagenetic changes to which it is subjected. In every case the isotopic composition of the final residue will largely reflect that of its precursor, although increases in $^{13}C$ content due to the preferred breakage of $^{12}C$–$^{12}C$ bonds and the loss of isotopically 'light' smaller molecules accompany metamorphic change.[10] If reducing conditions do not prevail during the initial deposition of the biogenic material, or do not persist during diagenesis, oxidation leads to the formation of carbon dioxide, correspondingly depleted in $^{13}C$, and the possible deposition of biogenic carbonates. $\delta^{13}C$ values of $-54‰$ have been reported for these.[30]

(*d*) Biogenic carbonates which result from the bacterial reduction of sulphates where the energy required for this conversion is derived from the oxidation of organic residues. Such carbonates are commonly associated with sulphur deposits and provide firm evidence of their mode of genesis.[31]

(*e*) The products of methane-producing bacteria, as at Saanich Inlet.

## 3 Sulphur

Although massive reserves of sulphur occur as dissolved sulphate in the oceans, in evaporite beds, in organic combination in shales and in localized areas as the result of geothermal or volcanic activity,[32] the sulphur available to non-marine plants and organisms is often very limited. The only direct mode of transport of sulphur from the oceans, the major available source, to the land is by airborne sulphate.[33] The measurements of several investigators suggest that the concentration of sulphur in the unpolluted atmosphere rarely exceeds $5 \mu g m^{-3}$ (ref. 34), and thus where other sources of sulphur are not present, the main contribution to the upper soil layers may depend largely on the small quantities of dissolved sulphate in precipitation. Over geological time, changes in land levels relative to the oceans have resulted in sulphur-rich marine deposits, *e.g.* evaporites and shales, being situated above current ocean levels, and the leaching of these may add very considerably to the sulphate content of fresh waters.[35] A recent study of the sulphur content of waters in the MacKenzie Basin vividly illustrates the variations in sulphur content which may arise within one system as the result of these many processes.[36] These findings, and more particularly those relating to the sulphur

30 W. A. Hodgson, *Geochim. Cosmochim. Acta*, 1966, **30**, 1223.
31 H. G. Thode, R. K. Wanless, and R. Wallouch, *Geochim. Cosmochim. Acta*, 1954, **5**, 286.
32 W. T. Holser and I. R. Kaplan, *Chem. Geol.*, 1966, **1**, 93.
33 W. W. Kellogg, R. D. Cadle, E. R. Allen, A. L. Lazrus, and E. A. Martell, *Science*, 1972, **175**, 587.
34 H. W. Georgii, *J. Geophys. Res.*, 1970, **75**, 2365; 'Air Pollution', ed. A. C. Stern, Academic Press, New York, 1968.
35 G. J. Blair, *J. Austral. Inst. Agric. Sci.*, 1972, **37**, 113.
36 R. Hitchon and H. R. Krouse, *Geochim. Cosmochim. Acta*, 1972, **36**, 1337.

content of the atmosphere,[37] have shown that the content and isotopic composition of this sulphur may be varied by relatively small alterations and additions to the environment.

In spite of the fact that (or perhaps because) these well-documented isotopic variabilities and instabilities exist, little interest has been shown in the $^{34}S$ contents of the organic materials which result from the metabolism of such sulphur sources, and concern has been almost entirely focused on those factors which control or cause the variations in the isotopic composition of the biologically available inorganic sulphur. This situation contrasts sharply with that for carbon, where equilibration between the higher concentration of carbon dioxide in the atmosphere and the dissolved carbon dioxide, bicarbonate, and carbonate in waters generally ensures an adequate supply of carbon for photosynthesis, and the availability of this element rarely becomes a significant question in environmental studies. Accordingly, disturbances in the established isotopic pattern are often only briefly sustained, and massive alterations to the available carbon in any natural system are required before significant isotopic variations are evident.

The processes responsible for the direct primary production of organically combined sulphur are the direct assimilation of sulphate by living plants and microbiological assimilatory processes in which, during the oxidation or reduction of sulphur species, organic sulphur compounds are synthesized and retained within cell structures. Measurements of the sulphur content of dried biological residues indicate that this may be as large as 3% in aquatic plants, but is customarily nearer to 1%.[38] Average values of 0.9 and 1.1% have been given for marine algae and animals, respectively,[39] and 0.6, 0.5, and 0.3% for experimentally grown and harvested bacteria, an algae, and a yeast, respectively.[40] These values alone suggest that the large reservoirs of sulphur which occur in organic combination in coals, petroleum, and other fossil biogenic residues, sometimes in concentrations as high as 20%,[41] do not have their origins in these primary reactions, but result from interactions between preserved organic residues and reactive reduced sulphur species during diagenesis. Although this has long been held to be the case,[42] the exact nature of these reactions is unknown. Isotopic evidence indicates hydrogen sulphide to be the source of the organic sulphur in Black Sea muds[43] and clearly demonstrates that the organic sulphur in the Californian Basins is not directly

[37] R. Shaw and H. R. Krouse, Air Pollution Control Assoc. Pacific N.W. Internat. Section. Calgary Nov. 1971; T. A. Rafter, *Bull. Volcanol.*, 1965, **28**, 12.

[38] V. L. Mekhtiyeva and R. C. Pankina, *Geokhimiya*, 1968, **6**, 739.

[39] I. R. Kaplan, K. O. Emery, and S. C. Rittenberg, *Geochim. Cosmochim. Acta*, 1963, **27**, 297.

[40] I. R. Kaplan and S. C. Rittenberg, *J. Gen. Microbiol.*, 1964, **34**, 195.

[41] T. A. Rafter, in 'Biochemistry of Sulfur Isotopes', Proceedings of a National Science Foundation Symposium, Yale University, April 12–14, 1962, ed. M. L. Jensen, National Science Foundation, New Haven, 1962, pp. 94–97.

[42] H. G. Thode, J. Monster, and H. R. Dunford, *Bull. Amer. Assoc. Petrol. Geol.*, 1958, **42**, 2619; H. G. Thode and J. Monster, *Amer. Assoc. Petrol. Geol.*, 1965, Mem. 4, 367.

[43] A. P. Vinogradov, V. A. Grinenko, and V. I. Ustinov, *Geochemistry (U.S.S.R.)*, 1962, 973.

derived from organic materials contributing to the basin sediments.[39] The cleavage of organic molecules to yield hydrogen sulphide is considered to contribute only 0.5% to the total hydrogen sulphide content of the Black Sea,[43] and the entire sulphur budget in both cases indicates the biological reduction of sulphate to be the most important stage in the sulphur cycle.

This Report is not directly concerned with the nature of these highly altered organic residues except for their impact on the environment; however, it is of particular interest that although they often show remarkably small variation in $^{13}C$ content within classes, major fluctuations in $^{34}S$ content are common and often provide considerable insight into the conditions of formation or deposition.[44] To some extent the same is true for plants and micro-organisms where direct assimilation of sulphate occurs. Aquatic plants, both freshwater and marine, preferentially metabolize the lighter isotope by from 0.0 to 4.4‰ (relative to dissolved sulphate) during growth.[38] A bacterium, a green alga, and a yeast similarly produced isotopic fractionations of −2.5, −1.4, and −2.8‰, respectively,[40] and the isotopic composition of animals and plants from the Californian Basins differs from that of the seawater sulphate by an average of 1.1‰.[39] Clearly, on the available evidence, assimilatory processes result in little more than minimal isotopic fractionation, and by thus reflecting the isotopic composition of growth media provide information on the environment of formation of biological specimens. Little application has been found for this relationship and, other than the values for primary biological products reported here, only cysteine from hair seems to have been analysed.[45]

Since the isotopic composition of biological materials appears to be closely controlled by that of the sulphate available for growth, the large quantity of data which has been compiled on the distribution and isotopic composition of sulphate in the atmosphere, precipitation, rivers, and fresh and saline waters is of inherent interest to the organic goechemist, as it virtually indicates the range of isotopic values likely to be found in organic materials. Several isotopic studies of the origins and concentrations of sulphur in the atmosphere and in precipitation[33,37,46] include evidence directly related to changes induced by further additions of sulphur compounds. From these studies it is clear that the gaseous products released during the combustion of fossil fuels have low $^{34}S$ contents and by dilution decrease the $^{34}S$ content of the atmosphere. Lakes (+1.9 to +8.9‰) and rivers (−20 to +20‰) have been shown to be of variable composition,[47] whereas sea water exhibits such a remarkable constancy that it may be employed as a secondary standard.

Differences in the isotopic composition of sulphur released from geothermal and volcanic sources have been demonstrated by the New Zealand group, with

44 H. G. Thode and C. E. Rees, *Endeavour*, 1970, **29**, 24; J. W. Smith and B. D. Batts, *Geochim. Cosmochim. Acta*, 1974, **38**, 121.
45 A. Szarbo, A. Tudge, J. Macnamara, and H. G. Thode, *Science*, 1950, **111**, 464.
46 G. Östlund, *Tellus*, 1959, **11**, 475; N. Nakai and M. L. Jensen, *Geochem. J.*, 1967, **1**, 199; G. Cortecci and A. Longinelli, *Earth Planet. Sci. Letters*, 1970, **8**, 36; B. D. Holt, A. G. Engelkemeir, and A. Venters, *Environ. Sci. and Technol.*, 1972, **6**, 338.
47 N. A. Yeremenko and R. G. Pankina, *Geochemistry* (*U.S.S.R.*), 1971, **45.**

average values of −11.7 and −1.2‰, respectively, being reported for New Zealand samples and corresponding values of −11.0 and −4.7‰, respectively, for those from New Guinea.[48] The average $^{34}S$ content of sulphur emitted from the White Island fumarole is 0.0‰,[49] whereas the sulphide and sulphate associated with the Wairekei geothermal bore water have $\delta^{34}S$* values of −12.6 to −19.4‰ and +5.0 to −1.0‰, respectively.[50]

Since the oxidation and reduction of sulphur compounds by biological and abiological processes and the equilibration and fractionation of the isotopes between the resulting species control the composition of the final products, many studies of such interactions have been made. The role of both types of reaction has been illustrated in an examination of the sulphur compounds in solfataras in Yellowstone National Park.[51] Here it was shown that the sulphur is produced by the abiological oxidation of hydrogen sulphide, and the sulphate by biological oxidation of sulphur. Some exceptions are also quoted. These conclusions are drawn from comparisons of the degree and direction of isotopic fractionation between sulphur species in natural systems and experimental values determined in the laboratory using specific micro-organisms under controlled conditions.

Thode and co-workers first demonstrated that a fractionation of the isotopes occurred during the bacterial reduction of sulphate and later concluded[52] that at 25 °C a maximum enrichment of 27‰ $^{32}S$ in the sulphide, relative to the sulphate, might be expected. Following this and other early investigations, Kaplan and Rittenberg[39,53] examined a number of systems in which sulphur compounds were metabolized, and they reported the maximum isotopic fractionations obtained under their experimental conditions; these are shown in Table 2.

More recently it has been shown that the oxidation of sulphur to sulphate by *Thiobacillus denitrificans* results in a change in $^{34}S$ content of less than 1‰;[54] the reduction of sulphite to sulphide by *Salmonella paratyphi A* gives maximum isotopic fractionations of −33.5 and −20.7‰ under anaerobic and aerobic conditions, respectively,[55] and the instantaneous fractionation by *Salmonella heidelberg* during this reaction may be −44‰ anaerobically.[56] An interesting example of a symbiotic reduction of sulphate by two *Clostridium* cultures A and B has also been described. Culture A reduces sulphate to sulphite and

* $\delta^{34}S‰ = \frac{[^{34}S/^{32}S]_{sample} - [^{34}S/^{32}S]_{standard}}{[^{34}S/^{32}S]_{standard}} \times 1000$ where the standard is sulphur as troilite in the Canyon Diablo meteorite.

[48] T. A. Rafter, I. R. Kaplan, and J. R. Hulston, *N.Z. J. Sci.*, 1960, **3**, 209.
[49] T. A. Rafter, S. H. Wilson, and B. W. Shilton, *N.Z. J. Sci.*, 1958, **1**, 154.
[50] T. A. Rafter, S. H. Wilson, and B. W. Shilton, *N.Z. J. Sci.*, 1958, **1**, 1.
[51] R. Schoen and R. O. Rye, *Science*, 1970, **170**, 1082.
[52] A. G. Harrison and H. G. Thode, *Trans. Faraday Soc.*, 1957, **53**, 1648.
[53] I. R. Kaplan and S. C. Rittenberg, in ref. 41.
[54] V. L. Mekhtiyeva, *Geochemistry* (*U.S.S.R.*), 1964, 26.
[55] H. R. Krouse, R. G. L. McCready, S. A. Husain, and J. N. Campbell, *Canad. J. Microbiol.*, 1967, **13**, 21.
[56] H. R. Krouse and A. Sasaki, *Canad. J. Microbiol.*, 1968, **14**, 417.

**Table 2** *Maximum fractionation measured in the metabolites formed by micro-organisms of the sulphur cycle under controlled conditions. All enrichments are given relative to the* $^{34}S/^{32}S$ *of the starting compounds*

| *Primary process* | *Organism* | *Starting substance* | *End product* | $^{34}S$‰ |
|---|---|---|---|---|
| Sulphate reduction | *D. desulfuricans* | $SO_4^{2-}$ | $H_2S$ | −46.0 |
| Sulphite reduction | *D. desulfuricans* | $SO_3^{2-}$ | $H_2S$ | −14.3 |
| Sulphite reduction | *S. cerevisiae* | $SO_3^{2-}$ | $H_2S$ | −41.0 |
| Sulphate assimilation | *E. coli*, *S. cerevisiae* | $SO_4^{2-}$ | Organic S | −2.8 |
| Cysteine hydrolysis | *P. vulgaris* | Cysteine | $H_2S$ | −5.1 |
| Chemosynthetic oxidation | *T. concetivorus* | $H_2S$ | S° | −2.5 |
| | | $H_2S$ | $SO_4^{2-}$ | −18.0 |
| | | $H_2S$ | $S_xO_y$ | +19.0 |
| Photosynthetic oxidation | *Chromatium* sp. | $H_2S$ | S° | −10.0 |
| | | $H_2S$ | $SO_4^{2-}$ | 0 |
| | | $H_2S$ | $S_xO_y$ | +11.2 |

culture B sulphite to sulphide, the maximum instantaneous fractionation in each case being 1.017 and 1.040, respectively.[57] The same authors also note that during the reduction of sulphite by other *Clostridium* species, the sulphide produced became progressively 'lighter' as the reaction proceeded, the inverse of the usually found isotope effect. Apart from this final item and one other previous report, all evidence appears to be in accord with the view[40] that all metabolic processes fractionate the isotopes of sulphur, other than those in which elemental sulphur is the starting material, and that the more reduced products or reactants are always enriched in $^{32}S$ relative to the starting material. The acceptance of this general principle allows a meaningful interpretation of isotopic data to be made, whether, for example, it be related to the oxidation of the organic sulphur in soils to sulphate, or the oxidation of sulphide to sulphur in oxygenated levels above anoxic basins.

**Biological Cycling.**—By combining isotopic data with the calculated quantities of sulphur occurring in various types of rocks, in solution in both oceans and rivers, and in the atmosphere, the main geochemical cycle of sulphur has been outlined.[32,42] The relatively minor role played by metamorphic and igneous rocks as a source of sulphur in the cycle is clearly demonstrated, in contrast to the importance attached to the aerial transport of sulphate from the oceans and, over much longer periods, the deposition of evaporites and shales. The major fractionation of the isotopes occurs during the bacterial reduction of ocean sulphate and although the isotopically 'light' sulphide so produced is preserved largely as pyrite, interactions with organic residues result in shales, crude oils, or coals with high organic sulphur contents. As stated previously, this organic sulphur as such is of little direct environmental interest as it is not related isotopically to primary biological products and is generally not available for biological utilization.

[57] V. Smejkal, F. D. Cook, and H. R. Krouse, *Geochim. Cosmochim. Acta*, 1971, **35**, 787.

More detailed attention is paid to the sulphur cycle in the atmosphere and the oceans in a more recent review[33] which is primarily concerned with differentiating between man-made and natural contributions of sulphur to these reservoirs. Four sources of sulphur to the atmosphere are considered: (*a*) man-made from the combustion of fossil fuels; (*b*) volcanic emissions; (*c*) sulphate in sea spray and biogenic marine hydrogen sulphide; and (*d*) hydrogen sulphide from biogenic processes on land. The importance of (*a*) in industrialized areas and the purely localized significance of (*b*) are not in dispute. However, the isotopic data previously presented indicate that sea-water sulphate and man-made contaminants are the major sources of atmospheric sulphur, a finding which is at variance with those of the above authors, which require a major contribution ($2.7 \times 10^8$ ton) of biogenic sulphide to the atmosphere. The question remains unresolved although studies of the atmospheric sulphur in and near Salt Lake City indicate that seasonal evolutions of bacteriogenic sulphur may be most significant.[58]

It is claimed that the sulphur-deficient areas of the world are increasing because of a general decrease in the availability of sulphur. Reasons given for this decrease are the increasing aversion to the combustion of sulphur-rich fuels in the interest of cleaner air, and an increasing need to economize in the application of sulphur-containing fertilizer. In this light, the forms of sulphur in soils and the atmosphere, the sources and variations in the supply of these, the ability of plants and animals to metabolize the available materials, and the transport of sulphur within plants have been discussed in some detail in a review[35] of the sulphur cycle in soil, plants, and animals. Since sulphur uptake by plants is almost exclusively through the root system as sulphate and the greater part of the sulphur in soils is in organic combination, it seems that the conversion of the organic sulphur into sulphate and the direct assimilation of this by the plant are important stages in the sulphur cycle. No isotopic data on the first of these stages have been reported.

## 4 Nitrogen

The recent controversy[59] regarding the value of $^{15}N/^{14}N$ ratio measurements in determining the source of the increasing concentrations of nitrate in the waters of Lake Decateur probably best illustrates the complexity of the nitrogen cycle in soils, the current lack of unequivocal data available, and the inherent difficulties in interpreting such data. Well-defined microbiological processes undoubtedly play a major role in the transport of nitrogen between the biosphere and the atmosphere but since (*a*) these processes may be accompanied by either a large isotopic fractionation or one of minor or negligible proportions, (*b*) the relative contributions made by these processes in

[58] D. C. Grey and M. L. Jensen, *Science*, 1972, **177**, 1099.

[59] D. H. Kohl, G. B. Shearer, and B. Commoner, *Science*, 1971, **174**, 1331; 1972, **177**, 454; R. D. Hauck, W. V. Bartholomew, J. M. Bremner, F. E. Broadbent, H. H. Cheng, A. P. Edwards, D. R. Keeney, J. O. Legg, S. R. Olsen, and L. K. Porter, *ibid.*, 1972, **177**, 453.

any system are not readily determined, and (*c*) a very considerable reservoir of nitrogen of varying isotopic composition is usually present in soils, the mechanism for nitrogen metabolism and transport in any system is not immediately evident from isotopic data.

Gaseous nitrogen in the atmosphere represents the major isotopically-invariant, natural reservoir of nitrogen available to land plants and organisms. However, Stevenson[60] quotes the assertion that the nitrogen dissolved in precipitation may reach values as high as 18.7 lb $acre^{-1}$ $year^{-1}$ and, since the same author reports that the biological fixation of nitrogen can scarcely provide more than another 50 lb $acre^{-1}$ $year^{-1}$ to soil, the contribution from the first source can be important locally. Although Hoering[61] has demonstrated that the isotopic composition of nitrate in rainwater varies from −0.1 to +9.0‰ relative to atmospheric nitrogen, little attention has been paid to this as a source of nitrogen or as a factor in influencing $^{15}N$ contents. Furthermore, it was shown that this nitrate results from the oxidation of ammonia of biogenic origin and not from the electrical fixation of atmospheric nitrogen. Since the greater part of atmospheric ammonia has a continental source,[62] this sequence may also be of considerable significance in the nitrogen cycle.

Better-documented microbiological processes include: (*a*) The fixation of atmospheric nitrogen by *Azotobacter*. Early experiments using four species of *Azotobacter* indicated that only one of these, *A. vinelandii*, produced even a slight fractionation of the isotopes, with the fixation of $^{14}N$ being favoured by 2‰.[63] More-recent studies have shown that under more-favourable conditions the fractionation factor may increase to 1.004.[64]

(*b*) The assimilation of ammonium by *A. vinelandii* and three soil yeasts, which results in an increase in the $^{15}N$ content of the residual ammonium due to the preferential utilization of $^{14}NH_4$. The fractionation factor for the bacterium was 1.015, and factors of 1.003 or less were found for the yeasts.

(*c*) Nitrification using *Nitrosomonas europaea*,[64] which resulted in an enrichment of the residual ammonium source in $^{15}N$ and the production of isotopically 'light' nitrite. A fractionation factor of 1.026 is reported.

(*d*) Denitrification with a wide variety of micro-organisms. Nitrogen depleted in $^{15}N$ by 17[64] and 20—30‰[65] is released on the reduction of nitrate by *Pseudomonas denitrificans* and *P. stutzeri*, respectively. Similar large kinetic isotope effects have been found with *Bacillus* and *Alcaligenes* species, and it is suggested that breakage of the N—O bond is not the total rate-controlling step in this conversion and that relatively stable intermediates tend to accumulate during the process.[65] This view is in accord with the findings of

[60] F. J. Stevenson, *Amer. Soc. Agronomists Monograph*, 1965, **10**, 1.
[61] T. C. Hoering, *Geochim. Cosmochim. Acta*, 1957, **12**, 97.
[62] S. Tsunogai, *Geochim. J.*, 1971, **5**, 57.
[63] T. C. Hoering and H. T. Ford, *J. Amer. Chem. Soc.*, 1960, **82**, 376.
[64] C. C. Delwiche and P. L. Steyn, *Environ. Sci. and Technol.*, 1970, **4**, 929.
[65] R. P. Wellman, F. D. Cook, and H. R. Krouse, *Science*, 1968, **161**, 269; F. D. Cook, R. P. Wellman, and H. R. Krouse, International Symposium on Hydrogeochemistry and Biogeochemistry, Tokyo, September, 1970.

Brown and Drury[66] who, on experimental and theoretical grounds, argue that a fractionation factor of 1.075 may be expected when cleavage of the N—O bond is totally rate-controlling.

Although a considerable range of $\delta^{15}N$* values has been reported for primary biogenic materials, *e.g.* land animals and products, +4.2 to +7.5‰; marine plankton, +3 to +13‰; seaweed, +8.1‰; clam flesh, +7.3‰; marine fish, +10 to +20‰; land plants, −6.5 to +6.2‰ and −2.2 to +5.0‰,[67] the total evidence strongly indicates that, in general, biologically combined nitrogen is enriched in $^{15}N$ relative to the atmosphere, exceptions being legumes, which are capable of fixing atmospheric nitrogen directly. From this, Parker[27] contends that denitrification processes, with the accompanying release of 'light' nitrogen into the atmosphere and the retention of $^{15}N$-enriched nitrate in the soil for subsequent biological utilization, probably largely control isotopic distribution.

Studies of the distribution and isotopic composition of nitrogen in soils[64,68] have revealed that the total nitrogen is generally enriched in $^{15}N$ relative to the atmosphere, $\delta^{15}N$ values of −1 to +17‰ are reported, and that total nitrogen contents usually correlate well with $^{15}N$ contents (possibly a result of the addition of 'heavy' fertilisers). However, even larger isotopic variations are noted when the various forms of nitogen are separated and examined. In one silty loam $\delta^{15}N$ values of +25‰ and +19‰ were found for hexosamine and hydroxy-amino-acids, respectively, and −3‰ for non-hydrolysable nitrogen.

Fossil fuels rarely contain more than 2% of nitrogen in organic combination, although considerable quantities of gaseous nitrogen may occur in association with natural gas, and the contribution of elemental nitrogen to the atmosphere which results from their combustion is negligible. However experimental data indicate that during combustion, particularly at lower temperatures, the chemically combined nitrogen in the fuel is converted into oxides of nitrogen, $(NO)_x$, more readily than the nitrogen entrained in the air required for combustion.[69] Since $\delta^{15}N$ values reported for fossil fuels range from −2.8 to +3.5‰ for coals and from +1.0 to +14.6‰ for crude oils, with a far greater range of values being given for oil gases and natural gas,[61,67,70] and since the concentration of $(NO)_x$ may reach almost 1 p.p.m.

* $\delta^{15}N‰ = \dfrac{[^{15}N/^{14}N]_{\text{Sample}} - [^{15}N/^{14}N]_{\text{Standard}}}{[^{15}N/^{14}N]_{\text{Standard}}} \times 1000$ where the standard is atmospheric nitrogen.

[66] L. L. Brown and J. S. Drury, *J. Chem. Phys.*, 1967, **46**, 2833; 1969, **51**, 3771.

[67] T. C. Hoering, *Science*, 1955, **122**, 1233; Y. Miyake and E. Wada, *Records of Oceanographic Works Japan*, 1967, **9**, 37; A. Parwel, R. Ryhage, and F. E. Wickman, *Geochim. Cosmochim. Acta*, 1957, **11**, 165.

[68] H. H. Cheng, J. M. Bremner, and A. P. Edwards, *Science*, 1964, **146**, 1574.

[69] D. W. Turner, R. L. Andrews, and C. W. Siegmund, *Combustion*, 1972, **44**, 21; G. B. Martin and E. E. Berkau, paper presented at meeting of the American Institute of Chemical Engineers, Atlanta City, August, 1971.

[70] R. Eichmann, A. Plate, W. Behrens, and H. Kroepelin, *Erdöl u. Kohle*, 1971, **24**, 2; C. Bokhoven and H. J. Theeuwen, *Nature*, 1966, **211**, 927; T. C. Hoering and H. E. Moore, *Geochim. Cosmochim. Acta*, 1957, **13**, 225.

in polluted atmospheres,[71] major variations in the isotopic composition of the oxidized nitrogen species in the atmosphere and in precipitation could arise from this cause. No isotopic data on this effect have yet been published.

The data available on the oceans indicate that solution of nitrogen gas is accompanied by a fractionation of the isotopes, with the dissolved gas being enriched in $^{15}N$ by 1‰. At ocean depths of 500 m or more, $\delta^{15}N$ values of +5 to +7‰ have been reported for both ammonia and nitrate, although one value for the ammonia in surface water showed a reverse effect of −3.5‰. Biogenic materials are enriched in $^{15}N$, with $^{15}N$ contents increasing along the possible food chain and with increasing biological and chemical complexity in the order: dissolved gaseous nitrogen < inorganic nitrogen forms < phytoplankton and algae < zooplankton and fish.[67]

**Biological Cycling.**—Measurements of the distribution of $^{15}N$ in Nature had until recently been so few that a meaningful interpretation of the data within the framework of the nitrogen cycle was not possible. However, the additional information which has become available now makes limited generalizations permissible[65] and, at the same time, places some restrictions on possible processes and mechanisms. For example, since it now seems to be agreed that the isotopic fractionation accompanying nitrogen fixation by soil bacteria in Nature is insignificant, the generally increased $^{15}N$ contents of biological materials must depend on other processes. Nitrate is usually available in rainwater, and thus Parker suggests that denitrification of nitrates to leave isotopically 'heavy' residues for biological utilization probably best explains this phenomenon. Similarly, if little isotopic fractionation accompanies the direct fixation of dissolved nitrogen by the marine algae *Trichodesmium* and other organisms, it seems that a denitrification process may equally well be required as the preliminary step to explain the formation of $^{15}N$-enriched marine residues.

The extent of man's impingement on the nitrogen cycle has not yet been investigated but the application of traditional fertilizers, and now those obtained by the fixation of atmospheric nitrogen, the release of fossil nitrogen as $(NO)_x$ during combustion, and the disposal of organic wastes in the oceans are probably not without measurable quantitative and isotopic effects. That difficulties exist in demonstrating these effects is apparent from the literature.

## 5 Hydrogen

The factors responsible for the distribution of the stable isotopes of hydrogen in the hydrosphere, the major reservoir of the element, have been described. Differences in the evaporative rates of the two oxides primarily control isotopic separation and thus both salinity and temperature strongly influence the extent of fractionation.[72]

71 J. Bogg, in 'Air Pollution Control', Part 1, ed. W. Strauss, Wiley, New York, 1971.

72 I. Friedman, *Geochim. Cosmochim. Acta*, 1953, **4**, 89; S. Epstein and T. Mayeda, *ibid.* p. 213; C. D. Garlick, in 'Handbook of Geochemistry', ed. K. H. Wedepohl, Springer-Verlag, Berlin, 1970 p. 8–B–1.

The ease of hydrogen exchange between water and hydrogen bonded to oxygen, nitrogen, and sulphur presents a problem in the measurement of D/H ratios in organic materials and, unless special precautions are taken, the resulting data may be more representative of a recent exchange product rather than the material of interest. Early measurements[73] on natural biogenic residues, *e.g.* coal and crude oils, were conflicting when compared with those for surface water, but more recent evidence has clearly shown that the lighter isotope is preferentially metabolized in biological processes.[74] Further experimental confirmation of the direction and extent of this effect resulted from measurements of the D contents of fatty acids from fish, rats, and a sediment core sample, and of the D contents of the body fluids of the animals and the water associated with the core. In all cases the acids were depleted in D by 21‰ relative to the corresponding water phase. This result has been interpreted as indicating that in each case the fatty acid was formed by the same mechanism, a mechanism which has not altered in $10^7$ years, the age of the sediment.[75]

Studies[76] of the isotopic composition of plants and of the moisture available for growth have shown that land plants are generally depleted in D by 30 to 40‰ relative to the water supply. No significant variation in the isotopic composition of the trunk, branches, leaves, and bark of a single plant was found, although the sap was enriched in D, presumably as the result of evapotranspiration. Marine specimens showed a greater isotopic variability, with codfish being depleted in D by 13‰, sea urchins and mussels by 82—99‰, seaweed by 57—103‰, and a moss-like alga by 168‰, relative to sea water. This general trend in $\delta$D* values has been confirmed by another detailed study of a collection of plants and animals using an essentially uniform source of sea water.[8] Despite large variations in $^{13}$C content ($-10$ to $-27$‰), possibly resulting from the utilization of different pathways for carbon fixation, no corresponding changes in D content were observed, and the plants were fairly constantly depleted in D by 55‰ relative to the water. Lipid fractions were further depleted in D by 90‰ relative to the whole plant. No isotopic variations between the organs of a plant, *e.g.* roots, leaves, were found. The D contents of the animals were generally less than those of the plants, with $\delta$D values of $-69$ and $-160$‰ being obtained for whole animals and lipid extracts, respectively.

Although the complete pattern of hydrogen metabolism in plants is not yet

$$* \ \delta D‰ = \frac{[D/H]_{Sample} - [D/H]_{Standard}}{[D/H]_{Standard}} \times 1000$$

Several isotopic standards have been employed; however, since most of the data reported here are comparative, the actual standard chosen is not significant. Where $\delta$D values are quoted they are relative to S.M.O.W.

73 K. Rankama, 'Progress in Isotope Geology', Interscience, New York, 1963, p. 153.

74 D. Bokhoven and H. H. J. Theeuwen, *Koninkl. Ned. Akad. Wetensschap. Amsterdam Proc. Ser. B*, 1956, **59**, 78.

75 G. Zborowski, L. Porticorvo, and D. Rittenberg, *Proc. Nat. Acad. Sci. U.S.A.*, 1967, **58**, 1660.

76 W. E. Schiegl and J. C. Vogel, *Earth and Planet. Sci. Letters*, 1970, **7**, 307.

fully understood, three stages resulting in isotopic fractionation have been suggested on the available information.

(*a*) Membrane discrimination between the isotopes is small and thus the separation of intracellular water from extracellular water is accompanied by a fractionation factor of 1.011.

(*b*) The reduction of carbon dioxide by hydrogen and the production of oxygen and carbohydrates results in the establishment of a marked fractionation pattern between the water and the photosynthetic products (fractionation factor 1.042).

(*c*) The synthesis of lipids from carbohydrates produces the greatest fractionation, and an average factor of 1.092 is proposed.

A method of estimating palaeotemperatures from corrected D/H ratios in peat has very recently been suggested.[77]

## 6 Oxygen

Although the effects of the biological cycling of oxygen are many and far-reaching, and perhaps the greatest attention has been paid to investigations of the isotopic distribution of $^{18}O$ in Nature,[78] no data on the isotopic composition of organically bound oxygen in natural products appear to exist. Early results suggested a preferential consumption of dissolved $^{16}O$ by marine organisms, and more recent data have confirmed the direction of this effect and shown that the higher $^{18}O$ content of atmospheric oxygen, +23‰, relative to sea water is not inconsistent with the effects to be expected from biological cycling. The two contributing processes are photosynthesis, which discriminates against $C^{16}O^{18}O$ and thus gives rise to the production of oxygen enriched in $^{16}O$, and respiration, which by operating more effectively in the reverse direction, and preferentially consuming $^{16}O$, increases the $^{18}O$ content of the atmosphere. The Dole Effect can be explained by calculating the net result of these processes from the various experimentally determined fractionation factors.[79] The role of biological processes in producing other apparently non-equilibrium situations has previously been discussed with regard to the $^{13}C$ content of the carbonates secreted by certain organisms. The same experimenters found similar non-equilibrium variations in the $^{18}O$ contents of the carbonates from these organisms.[28,29]

Whilst the $^{18}O/^{16}O$ ratio in sulphate from the shells of living molluscs is not significantly different from that of the sulphate dissolved in the water in which they grow,[80] the bacterial reduction of sulphate can produce variations in $^{18}O$ contents. Mizutani and Rafter report that at any stage in the bacterial

77 W. E. Schiegl, *Science*, 1972, **175**, 512.

78 S. R. Silverman, *Geochim. Cosmochim. Acta*, 1951, **2**, 26; S. Epstein, in 'Researches in Geochemistry', ed. P. Abelson, Wiley, New York, 1959 p. 217.

79 M. Dole and G. Jenks, *Science*, 1944, **100**, 409; N. M. Rakestraw, D. P. Rudd, and M. Dole, *J. Amer. Chem. Soc.*, 1951, **73**, 2976; G. A. Lane and M. Dole, *Science*, 1956, **123**, 574.

80 G. Cortecci and A. Longinelli, *Earth and Planet. Sci. Letters*, 1971, **11**, 273.

reduction of a limited reservoir of sulphate, the residual sulphate is enriched in both $^{18}O$ and $^{34}S$, relative to the starting material, the ratio of these enrichments being 1:4.[81]

It seems that the present understanding of biological oxygen cycles is so limited that the interpretation of data relating to the environment will be problematical. Presumably the huge oxygen demand of the vast quantities of organic material which are injected into the oceans at localized points will produce a measurable isotopic effect. No observations on this have as yet been reported.

## 7 General Conclusions

With the exception of carbon, relatively few measurements have been made of the isotopic composition of the light elements in naturally occurring organic compounds. However, the gradual accumulation of information on the distribution and isotopic composition of all of these elements allows an increasing reliance to be placed on the interpretation of data, and is revealing the presence and effects of natural processes which previously had not been recognized. The awareness of these processes and the increasing realization that many major reservoirs of carbon are of variable isotopic composition has resulted in less emphasis being placed on absolute isotopic ratios and a greater regard being paid to variations and comparisons within well-defined systems.

Whilst isotopic measurements will continue to be used extensively in determining the origins and fate of organic materials depositing under sedimentary conditions, increasing population pressures are demanding that, in the interests of greater food production and cleaner air, attention be directed to gaining a better understanding of the nitrogen and sulphur cycles, and the sources of these elements and their compounds, in soils and the atmosphere. Isotopic ratios promise to be of considerable aid in these respects and already the relative ease with which major sources of sulphur pollutants in the atmosphere may be determined ensures a continuing endeavour in this direction. Similarly, the role of biogenic sulphide in the atmospheric sulphur cycle remains to be unequivocally demonstrated. Confirmation of the scant isotopic data on nitrogen compounds in the atmosphere is also required. From this it appears that elemental nitrogen is rarely the source of these and that nitrate results from the oxidation of biogenic ammonia of continental origin. Recent experiments also indicate that the nitrogen oxides produced during the combustion of fossil fuels originate from the organic nitrogen compounds in these fuels and suggest valuable diagnostic application of $^{15}N/^{14}N$ ratio studies.

Whether the quantities of nitrate so produced are sufficient to alter significantly the isotopic composition of local soils or plants is unknown. Sulphate

[81] Y. Mizutani and T. A. Rafter, *N.Z. J. Sci.*, 1968, **12**, 60.

carried in precipitation from the same source may similarly have an effect on local plant growth owing largely to its ease of assimilation. No measurements of the $^{34}S/^{32}S$ ratios of the organically bound sulphur in plants existing in polluted situations appear to have been made.

The extremely limited data available on the biogeochemical cycling of hydrogen and oxygen and on the distribution of D and $^{18}O$ in natural organic products have curtailed attempts to apply the measurement of these to solving environmental problems in organic chemistry. One immediate future application appears to be the determination of changes in $^{18}O$ content which accompany the discharge of organic wastes into continental seas.

# 2

# Environmental Organic Chemistry of Rivers and Lakes, Both Water and Sediment

BY P. A. CRANWELL

## 1 Introduction

Interest in the organic chemistry of the freshwater environment has increased in the past decade, due in part to advances in analytical techniques, enabling identification of compounds present at the sub-microgram level, and in part from concern for those aquatic habitats under pollution stress. The major aspects of research concern:

(1) Water supply—the necessity to produce potable water supplies from increasingly contaminated natural waters has stimulated attempts to isolate and identify the organic compounds present in water as an aid to treatment and purification.

(2) Pollution by toxic chemicals—concern with the fate and residence time of pesticides in the aquatic environment has led to the development of analytical techniques suited to the study of their concentration in the food chain and of the metabolites resulting from decomposition.

(3) Effects of nutrient enrichment, which results in large algal populations in standing waters and an increased rate of accumulation of organic material in the sediments. Organic chemists interested in the pathway of carbon in Nature have begun to study the role of micro-organisms and the effect of aerobic or anaerobic conditions on the early stages of diagenesis of organic matter in sediments.

(4) The realization that lake sediments represent a chronological sequence of deposition of partly organic material derived from the lake and its drainage basin and that preservation of organic matter depends both on whether aerobic or anaerobic processes prevail at the sediment—water interface and on the relative rates of addition and utilization, has led to chemical studies on the sediment profile in order to interpret past conditions of the lake basin from the nature of the sediments.[1]

**Economic Significance.**—Rivers and lakes, natural and man-made, have economic significance in water storage and irrigation, in provision of power

[1] D. G. Frey, *Mitt. int. Verein. theor. angew. Limnol.*, 1969, **17**, 7.

for industry, for amenity and recreational use, and also for commercial fisheries.

**Nature of the Freshwater Environment.**—The chemical nature of water bodies is determined by the interaction of geographic, geologic, physical, and biological factors. Freshwater habitats can be divided into two groups: (1) standing water, such as lakes and ponds, in which nutrients may cycle several times, and (2) running water where, in contrast to (1), there is permanent removal of eroded and transported material from a given point. The suspended material in running water is deposited mainly as sediment in a lake along its course or, on reaching the sea, in an estuary. Since the estuarine environment is discussed elsewhere, and river sediments have received little attention, further discussion will mainly concern river and lake waters and lake sediments.

Lakes are usually classified according to either their nutrient content or the rate of organic matter production; thus oligotrophic lakes are low in dissolved nutrients such as phosphorus, nitrogen, and calcium, since they lie on infertile rock. They are usually deep, having a small littoral zone and are situated in mountainous areas. Because of low productivity, oxygen depletion of the hypolimnion (lower layer of water in a thermally stratified lake) does not occur in this type of lake. Eutrophic lakes (nutrient-rich) lie on or receive drainage from rocks rich in plant nutrient elements, they are usually shallow, with a large littoral area, and show oxygen depletion of the hypolimnion in summer owing to the decomposition of plant and animal remains by bacteria. Thus eutrophication is defined as enrichment by nutrients or organic matter, or both, that results in high biological productivity and a decreased volume within the ecosystem.[2] In undisturbed lakes eutrophication is a natural ageing process, leading to the eventual disappearance of the lake itself, which can be accelerated by man either by discharge of nutrients in the form of industrial and domestic waste or by agricultural practices, such as land drainage or deforestation, which result in increased erosion of the drainage basin.

The concentration of dissolved oxygen in natural waters affects their chemical nature. During periods of thermal stratification, the hypolimnion of a lake remains out of contact with the atmosphere and may become deoxygenated, as outlined above. The reducing conditions aid preservation of labile organic compounds such as chlorophyll and its degradation products. Solubilization of inorganic species such as iron by reduction of $Fe^{3+}$ to $Fe^{2+}$ also leads to mobilization of phosphate, thus producing an increased nutrient supply.[2] The relationship between oxidation–reduction potential ($E_H$) and oxygen concentration across the sediment–water interface has been examined.[3]

[2] G. E. Likens in 'Nutrients and Eutrophication', Special Symposium, American Society for Limnology and Oceanography, Lawrence, 1972, pp. 3—13.
[3] B. T. Hargrave, *Oikos*, 1972, **23**, 167.

## 2 Sources of Organic Matter

Organic matter inputs to the aquatic ecosystem may originate from allochthonous or autochthonous sources. The former term refers to biological residues of non-aquatic origin. Leaching of terrestrial vegetation and soils of the drainage basin and erosion due to rapid run-off of water result in the transport of dissolved and particulate matter downstream, while standing water is a final resting place for wind-borne leaves, which have been described as meteorologic input.[2] Autochthonous inputs include carbon fixed within the aquatic environment and may be sub-divided into phytoplankton (microscopic plants floating in water), periphyton (plant organisms, except rooted macrophytes, growing on submerged materials in water), macrophytes (non-microscopic vegetation), and chemosynthetic micro-organisms. The relative importance of these sources varies in different environments, phytoplankton providing the main source of autochthonous material in large deep lakes, but periphyton and macrophytes becoming more important in shallow lakes and streams. In shaded or polluted streams autochthonous inputs may be negligible. Approximate annual rates of phytoplankton production in oligotrophic and naturally eutrophic lakes lie in the range 7–25 g carbon $m^{-2}$ $year^{-1}$ and 75–250 g carbon $m^{-2}$ $year^{-1}$, respectively, while culturally polluted lakes (those in which eutrophication has been accelerated by man) range between 350 and 700 g carbon $m^{-2}$ $year^{-1}$. Some of the fixed carbon is rapidly excreted as dissolved organic matter and will be discussed below.

Man-made organic compounds are introduced into the aquatic environment either directly, as industrial and sewage effluents and aquatic herbicides, or indirectly by the use of pesticides within the drainage area. Although effluents from specific industrial processes may contain naturally occurring compounds at unnaturally high concentrations, thus making difficult a clear distinction between natural products and some pollutants, reference to such examples has been avoided, so that obvious pollutants will be discussed in Section 6. Sections 3–5 review recent studies on the naturally occurring compounds in rivers and lakes, with special reference to the influence of trophic status on the quantity, diversity, and reactions of the compounds.

## 3 Organic Matter in Water

Organic matter in water has been sub-divided into dissolved and particulate fractions. Dissolved material is defined as that material which passes through some type of filter having a pore size of about 0.5 $\mu$m. By this definition living organisms are included in the particulate fraction and some colloidal material is included in the dissolved fraction. Some investigations on dissolved matter have been made on unfiltered samples, so that particulate matter will have been included in these analyses.

**Particulate Fraction.**—In a review of compounds identified in natural waters, sediments, sewage, and soils, Vallentyne[4] stated that there had been little investigation into the detailed composition of particulate matter, which constitutes less than 10% of the organic carbon in normal lakes and rivers but up to 60% in turbid rivers, corresponding to concentrations of particulate carbon up to 10 mg $l^{-1}$. The only compounds identified were carbohydrates obtained by hydrolysis of cellular material. More recent studies concerned with the release of dissolved organic material by bacterial attack on algae have reported the yield of the residual refractory material as a measure of the amount of carbon being added to the sediments. Bacterial attack on a wide range of algae for 1 year gave a resistant fraction of 12–86%.[5] The sorption of polynuclear aromatic hydrocarbons on suspended solids in surface waters has also been reported, but it is uncertain whether or not these are of industrial origin. Amino-acids bound to clay minerals were detected in the particulate fraction from a German river.[6]

**Dissolved Organic Matter.**—The review of Vallentyne[4] has been supplemented by Croll, who included pollutants in his survey[7] and also discussed methods for recovery of a wide spectrum of compounds from water.

*Simple Lipids.* There have been few recent reports of long-chain hydrocarbons and fatty acids in water. Solubility data recently reviewed[8] give values below $10^{-7}$ mol $l^{-1}$ for alkanes above $C_{14}$ and less than $10^{-5}$ mol $l^{-1}$ for fatty acids above $C_{14}$ at pH 5.7. Since it has been shown that, for micro-organisms in contact with fatty-acid solutions, intracellular concentrations of fatty acid are determined by the external concentration of undissociated fatty acid,[8] these solubilities are important when considering microbial breakdown. However, natural aquatic systems contain other organic components which may affect the solubility since it has been shown that fulvic acid, a water-soluble fraction of soil-organic matter, can solubilize long-chain hydrocarbons and fatty acids.[9] It is noteworthy that the n-alkanes in fulvic acid showed no odd-carbon predominance typical of higher plants, suggesting that they are of microbiological origin.

*Carbohydrates.* The polysaccharide fraction of a small productive lake gave galactose (1), glucose (2), mannose (3), arabinose (4), xylose (5), rhamnose (6), fucose (7), uronic acids, glucosamine (2-aminoglucose), and 2-deoxy-D-ribose (8). Only small amounts of mono- and oligo-saccharides were present.[10]

[4] J. R. Vallentyne, *J. Fisheries Res. Board Canada*, 1957, **14**, 33.
[5] W. J. Jewell and P. L. McCarty, *Environ. Sci. Technol.*, 1971, **5**, 1023; E. G. Foree and P. L. McCarty, *ibid.*, 1970, **4**, 842.
[6] A. A. Prashnowsky, G. Ebhardt, and M. Hobler in 'Advances in Organic Geochemistry 1971', ed. H. R. von Gaertner and H. Wehner, Pergamon Press, Oxford, 1972, pp. 403—413.
[7] B. T. Croll, *Water Treatment and Examination*, 1972, **21**, 213.
[8] G. H. Bell, *Chem. and Phys. Lipids*, 1973, **10**, 1.
[9] G. Ogner and M. Schnitzer, *Geochim. Cosmochim. Acta*, 1970, **34**, 921.
[10] G. Weinmann, *Arch. Hydrobiol., Suppl.* **37**, 1970, 164.

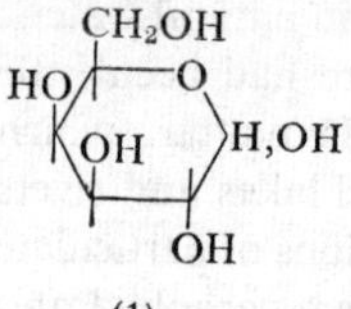

(1)

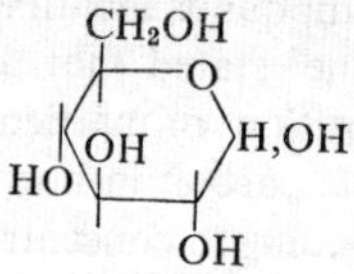

(2)

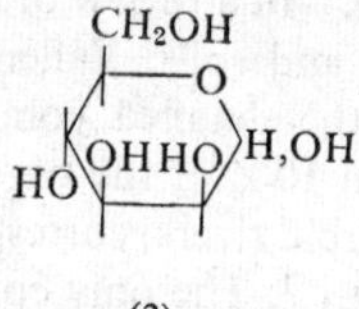

(3)

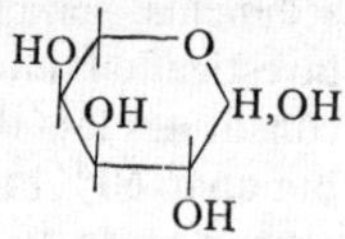

(4)

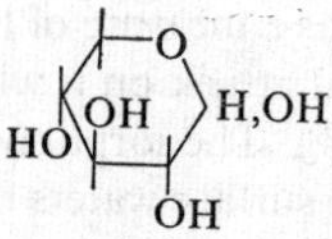

(5)

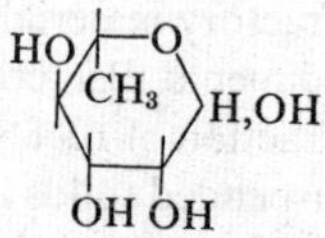

(6)

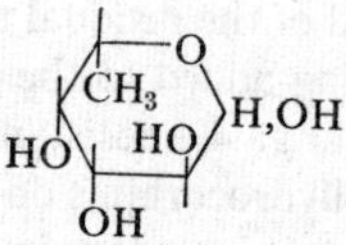

(7)

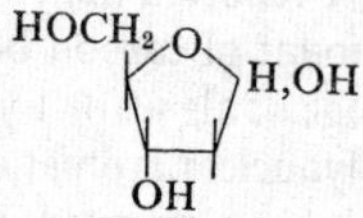

(8)

The same monosaccharides were found in an alga *Scenedesmus quadricauda* with and without bacteria. The turnover rate of glucose in a lake, determined by measurement of the uptake kinetics of $^{14}C$-labelled glucose by bacteria, was found to vary from 10 h in summer to 1000 h in winter. It was suggested that turnover rates provided a better indication of ecological importance than the steady-state concentration, which never exceeded 6 $\mu g\ l^{-1}$.[11]

*Organic Nitrogen Compounds.* Vallentyne[4] and Croll[7] have reviewed the amino-acids detected in hydrolysates of natural waters, and the content of free amino-acids. Recent work has concentrated on differences in content and composition of amino-compounds in waters of different trophic status. During comparative studies of a productive and less productive lake, Gocke[12] found that the dissolved amino-compounds constituted 5.5% and 10.3% respectively of the dissolved organic matter and constituted 23% and 32% respectively of the total dissolved organic nitrogen. The free amino-acids, dominated by serine (9; R = $HOCH_2$) and glycine (9; R = H), totalled 61

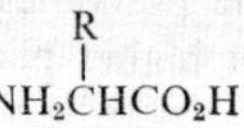

(9)

and 39 $\mu g\ l^{-1}$ in the respective productive and less-productive lakes while peptides totalled 250 and 226 $\mu g\ l^{-1}$, respectively. Glucosamine was also present in the two waters at concentrations of 126 and 31 $\mu g\ l^{-1}$, respectively (15% and 7% of the dissolved amino-compounds). Free amino-acids in natural waters have been isolated by a combination of ligand-exchange and ion-exchange chromatography and analysed by g.l.c. after conversion into the *N*-(trifluoroacetyl)methyl ester. Recovery of most of the standard amino-acids lay in the range 53—93%.[13]

[11] J. E. Hobbie in 'Chemical Environment in the Aquatic Habitat', ed. H. L. Golterman and R. S. Clymo, North Holland, Amsterdam, 1967, pp. 245—251.
[12] K. Gocke, *Arch. Hydrobiol.*, 1970, **67**, 285.
[13] W. S. Gardner and G. F. Lee, *Environ. Sci. Technol.*, 1973, **7**, 719.

Evidence concerning the effect of bacteria on the amino-acid composition was obtained by Gocke[12] from a comparison of a bacteria-free algal culture with a similar culture contaminated with bacteria. The amino-acid pattern differed qualitatively and quantitatively. In the pure culture aspartic acid (9; $R = HO_2CCH_2$) and glutamic acid (9; $R = HO_2CCH_2CH_2$) were dominant, whereas the contaminated culture was dominated by serine and glycine, as in the lakes examined.

Sources of amino-acids include excreta of living organisms, decomposition of the proteins of dead organisms within the water body, decomposition of dead leaves from terrestrial vegetation, and discharge of domestic waste water.[14]

Since the amino-acids in water are of biological origin they should be largely in the L-configuration; however, a small amount of racemization may occur during the residence time, which is believed to be governed by the rate of removal by organisms. There are few investigations of the rate of utilization of amino-acids in natural waters, though it has been suggested that in shallow lakes, where amino-acids and bacteria are fairly uniformly distributed in the water column, the turnover rate is fast at all depths.[15] Non-biological decomposition processes of amino-acids in natural waters have also been discussed.[15] Under aerobic conditions, the fastest reaction is a metal-ion-catalysed oxidation to form the corresponding keto-acid. The metal ion functions by chelation, and approximate half-lives of 1—7 h at pH 9.6 and 5—17 h at pH 4 were obtained for a range of amino-acids catalysed by $Cu^{2+}$. Estimation of the rate of this reaction in natural waters is difficult because of the reduced catalytic effect of the metal in the presence of natural chelators; however, calculations suggest that oxidation by this process would be insignificant where amino-acids have short residence times.

*Vitamins.* Several reports have noted the presence in natural waters of vitamins $B_1$ (10), $B_{12}$ (11), and biotin (12), which are essential for the growth of various freshwater algae and are detected by bioassay techniques. In a recently formed artificial lake, during periods of water circulation, concentrations of 40, 1, and 10 ng $l^{-1}$, respectively, were found, but during summer stratification the fluctuations in the vertical distribution corresponded with the succession of phytoplankton, and vitamin $B_{12}$ was believed to be produced by the activity of micro-organisms in sediments.[16] Two groups have reported a much greater concentration of these vitamins in eutrophic compared to oligotrophic lakes. In the latter type, vitamins were found only in the euphotic zone (the upper portion of the water column, in which photosynthesis occurs), with maximum concentrations of 1.1 ng $l^{-1}$, 38 ng $l^{-1}$, and 1.7 ng $l^{-1}$ of vitamins $B_{12}$, $B_1$, and biotin, respectively, whereas autumnal levels in a eutrophic lake were

[14] J. V. Hunter in 'Organic Compounds in Aquatic Environments', ed. S. D. Faust and J. V. Hunter, Dekker, New York, 1971, pp. 51—94.
[15] J. L. Bada in 'Nonequilibrium Systems in Natural Water Chemistry', Advances in Chemistry Series No. 106, Amer. Chem. Soc., Washington, 1971, pp. 309—331.
[16] K. Ohwada and N. Taga, *Limnol. Oceanogr.*, 1972, **17**, 315.

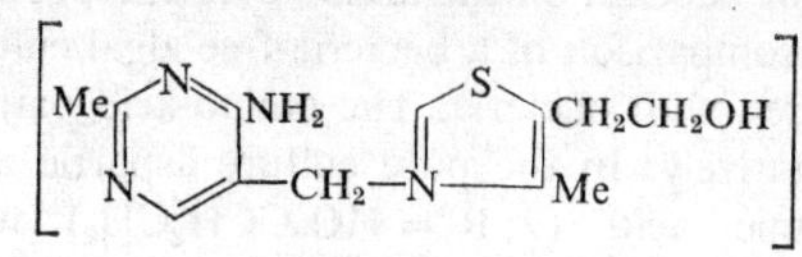

(10)

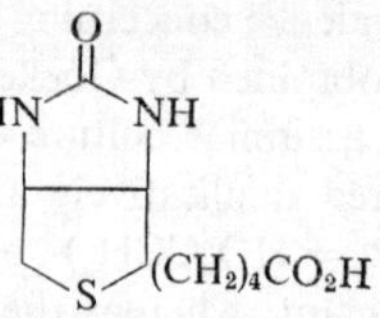

(12)

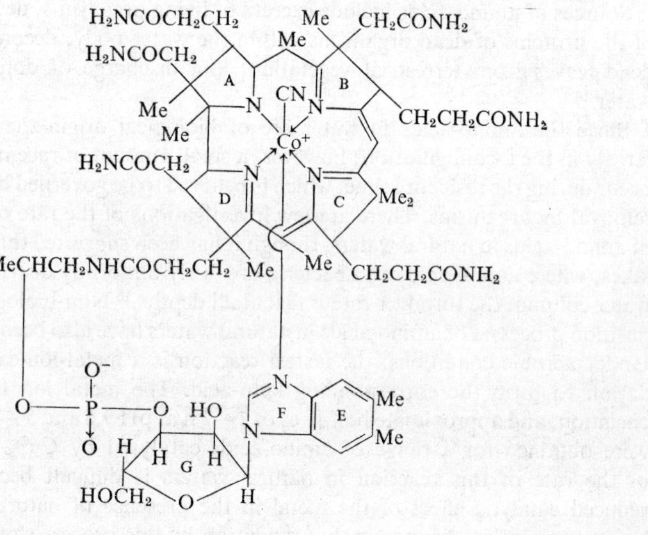

(11)

4.5, 400, and 6.4 ng $l^{-1}$, respectively.[17] Daisley had previously reported that the average annual content of vitamin $B_{12}$ in lakes of N.W. England correlated with the evolutionary series of the lakes, being low (1—2 ng $l^{-1}$) in unproductive and higher (10—15 ng $l^{-1}$) in productive lakes, which showed considerable seasonal variation in the quantity detected.[18]

*Compounds Responsible for Odours in Waters.* Recent developments in the chemistry of odour-producing compounds in water include the identification of 2-methylisoborneol (13) and geosmin (14) from a carbon-filter extract of

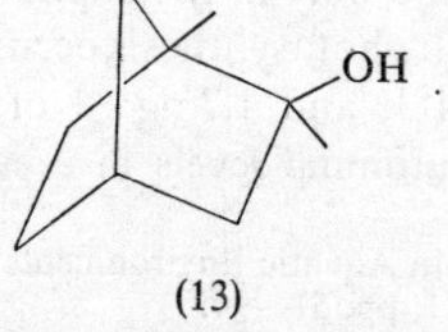

(13)

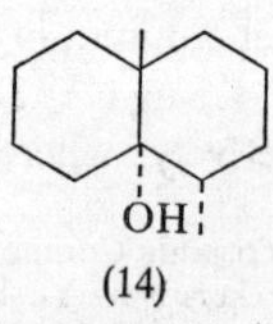

(14)

[17] A. F. Carlucci and P. M. Bowes, *Limnol. Oceanogr.*, 1972, **17**, 774.
[18] K. W. Daisley, *Limnol. Oceanogr.*, 1969, **14**, 224.

Grand Lake (Ohio).[19] These compounds are produced by actinomycetes and algae.

*Coloured Organic Substances.* The yellow or brown organic materials collectively called humic substances have been widely investigated and their phenolic and acidic natures established, but their specific structure remains unknown. They form the largest organic component in highly coloured natural waters, at concentrations of 10—30 mg $l^{-1}$ as carbon, and have been divided into humic, hymatomelanic, and fulvic acids according to their relative solubility in acid and alcohol. Isolation methods and techniques for determining the molecular size distribution have been reviewed.[7,20] There is an inverse relationship between the colour of a water and its productivity, so that metabolic by-products of aquatic organisms are not significant sources of the coloured material. Soil organic matter is suggested as the probable source, based on similarity of degradation products.[20] The variation in chemical properties of coloured substances from different waters suggests that the compounds in a particular water reflect the surrounding environment, but the nature of this relationshp has not been established. It is believed that these water-soluble humic compounds, which can solubilize metal ions by chelation, are important in controlling the supply and availability of metals needed for growth of algae.[21] A published method for determination of the organic C in water, by conversion into methane which is measured by a flame ionization detector, has been improved to allow analysis of natural waters containing $<1$ mg $l^{-1}$ organic carbon.[22]

*Release of Dissolved Organic Material.* The contribution to lakes and rivers of dissolved organic compounds from autochthonous and allochthonous sources has received considerable attention. Many investigations on autochthonous material concern the extracellular products secreted by algae. The literature was reviewed by Fogg,[23] who noted that extracellular release of organic compounds could provide a pathway by which organic matter produced by photosynthesis became available to non-photosynthetic organisms. One of the major products liberated is glycollic acid (15), which has been found in Lake Windermere[23] in concentrations up to 60 $\mu$g $l^{-1}$ and also in a productive lake at concentrations of 0.2—0.4 mg $l^{-1}$.[10] It is believed to be derived from a $C_2$ fragment produced by cleavage of one of the sugar-diphosphates of the carbon-fixation cycle, and the pathway of further metabolism appears to be *via* oxidation to glyoxylate then through glycine, serine, and glycerate to phosphoenol pyruvate and the tricarboxylic acid cycle.

[19] A. A. Rosen, C. I. Mashni, and R. S. Safferman, *Water Treatment and Examination*, 1970, **19**, 106.

[20] R. F. Christman and R. A. Minear in 'Organic Compounds in Aquatic Environments', ed. S. D. Faust and J. V. Hunter, Dekker, New York, 1971, pp. 119—143.

[21] M. Schnitzer in 'Organic Compounds in Aquatic Environments', ed. S. D. Faust and J. V. Hunter, Dekker, New York, 1971, pp. 297—315.

[22] B. T. Croll, *Chem. and Ind.*, 1972, 386.

[23] G. E. Fogg, *Arch. Hydrobiol., Beih., Ergeb. Limnol.*, 1971, **5**, 1.

Liberation of organic acids from phytoplankton has been reported[23] and may be the origin of short-chain acids such as formic, propionic, butyric, oxalic, malic (16), citric (17), and lactic (18) acids reported in a productive lake in

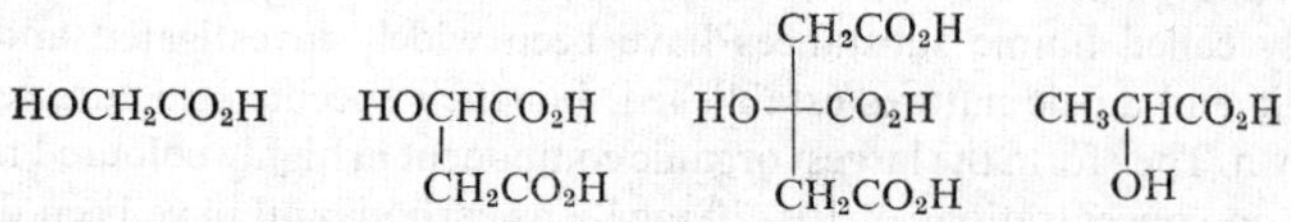

(15) (16) (17) (18)

addition to glycollic acid.[10] It was found that the concentration of extracellular compounds in a bacteria-free culture was greatly reduced in the presence of a mixed population of bacteria isolated from the lake in which these acids were found. Other extracellular algal products whose occurrence in natural waters has been discussed are polysaccharides, amino-acids, and polypeptides. The excretion of unidentified phosphorus-containing organic matter of low molecular weight by a natural phytoplankton population, followed by extracellular formation of a colloid containing phosphorus, has recently been reported.[24]

Studies on the secretion of dissolved material by macrophytes have shown that the littoral flora is a potential major source of dissolved organic matter (DOM) in lakes. The major compounds secreted by a submerged angiosperm, *Najas flexilis*, were found to be glucose (2), sucrose (19), fructose (20),

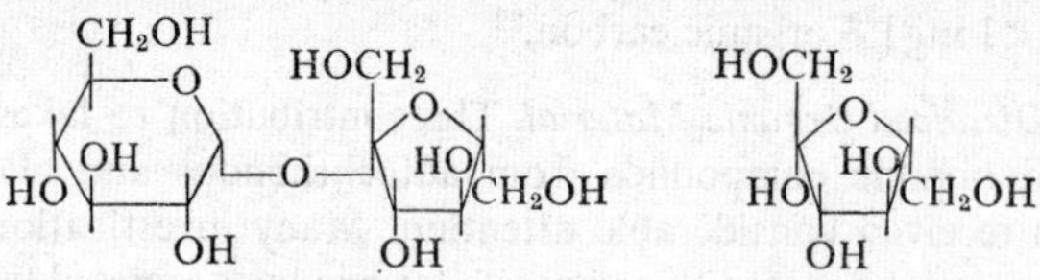

(19) (20)

xylose (5), and glycine (9; R = H).[25] During experiments to investigate the rate at which components were leached from decaying leaves in a simulated hard-water stream, a tenfold increase in dissolved organic carbon occurred in 30 h and the bacterial population which developed then rapidly decomposed the dissolved material within 72 h. Two fractions of the dissolved organic carbon were recognized, a bacteriologically labile component with half life ($t_{1/2}$) 2 days and a refractory component with $t_{1/2}$ 80 days, while a refractory organic nitrogen fraction was unmodified for 24 days. It was concluded that the dissolved organic-matter flux was controlled by bacterial utilization in streams as in lakes.[26]

Using a u.v. combustion technique (a more sensitive method for nitrogen than the micro-Kjeldahl) to distinguish between labile and refractory dissolved organic nitrogen fractions of natural lake waters differing in trophic

[24] D. R. S. Lean, *Science*, 1973, **179**, 678.
[25] R. G. Wetzel and B. A. Manny, *Verh. int. Verein. theor. angew. Limnol.*, 1972, **18**, 162.
[26] R. G. Wetzel and B. A. Manny, *Limnol. Oceanogr.*, 1972, **17**, 927.

status, it was shown that a eutrophic lake contained 5—10 times more labile dissolved organic nitrogen than an oligotrophic lake.[27]

Production of dissolved organic matter by bacterial attack on dead algal cells (*Scenedesmus* sp.), using a mixed culture obtained from a lake sediment, thus simulating decomposition at the sediment surface, has been studied under aerobic[28] and anaerobic[29] conditions. Under aerobic conditions the production of dissolved organic matter was about 7% as carbon and 6% as nitrogen after 30 days, and a proteinaceous compound, giving 14 amino-acids on hydrolysis, was recognized in the dissolved fraction. The decomposition rate during this time followed approximate first-order kinetics, and the decomposition pattern of the cell nitrogen suggested that the algal residue could be divided into labile and refractory constituents according to their resistance to bacterial attack. Under anaerobic conditions 30% of algal cell carbon and 8% of the nitrogen was dissolved after 60 days; $C_1$—$C_3$ carboxylic acids were found in solution and proteinaceous material was also obtained. Under anaerobic conditions the production rate of dissolved organic material was four times faster and the rate constant of decomposition of algal cell C and N decreased to less than half compared with aerobic decomposition. During a similar study[5] of the decomposition of axenic cultures and mixed cultures of algae, bacteria, and zooplankton it was found that the rate and extent of conversion of algae into soluble forms was essentially the same under aerobic and anaerobic conditions ($K = 0.01$—$0.03$ day$^{-1}$), in contradiction of the general belief that large accumulations of organic materials resulted from anaerobic conditions.

The kinetics of extracellular release of soluble organic matter by planktonic algae have been examined (using $^{14}C$) under conditions approximating to natural changes in light intensity, and the relative kinetics of cumulative net particulate carbon fixation and of extracellular release have been used to infer the general nature of the substrate being released.[30] The sequence: inorganic C $\rightarrow$ algae $\rightarrow$ soluble organic C $\rightarrow$ bacteria represents a catenary reaction sequence in which the time delay before later members are formed depends on the distance of each component from the starting point. In an oligotrophic subalpine lake a close superposition of photosynthesis and extracellular release was observed, whereas in a eutrophic lake a lag time of several hours was observed. The data suggested a continuous spectrum of release patterns, ranging from simple organic molecules to polymers and from release throughout the day to release only during daylight.

## 4 Organic Matter in Sediments

The quantity of organic matter carried in natural waters is small compared with that in sediments, which form the main deposits of organic debris

[27] B. A. Manny, *Verh. int. Verein. theor. angew. Limnol.*, 1972, **18**, 147.
[28] A. Otsuki and T. Hanya, *Limnol. Oceanogr.*, 1972, **17**, 248.
[29] A. Otsuki and T. Hanya, *Limnol. Oceanogr.*, 1972, **17**, 258.
[30] G. W. Saunders, *Verh. int. Verein. theor. angew. Limnol.*, 1972, **18**, 140.

accumulated during the earth's history. Sediment-trap studies showed that at least 70—90% of the organic matter synthesized in a column of lake water is decomposed before incorporation into the surface sediment, where further breakdown occurs.[31] Accumulation of sediment derived either from decaying cellular material or inorganic material washed into the lake is an ageing process which will fill the lake over a geological time-period. Natural eutrophication accelerates when the lake is sufficiently shallow to allow frequent complete circulation of the water column, so that nutrients are more rapidly recycled from the sediment. It has been observed that, despite wide variations in lake size and depth, in delivery of clastic and organic sediments from the drainage basin, and in human disturbance, the total thickness of organic sediments formed over approx. 10 000 years in a large number of lakes lies between 6 and 12 m.[32] Water bodies receiving nutrient enrichment resulting from the influence of man show an increased rate of sediment accumulation in the surficial sediment.[33]

The nature of the organic matter in sediments was the subject of a review in 1957.[4] During the past decade, advances in our knowledge of the nature of organic matter in the geosphere have resulted from the use of modern analytical techniques, such as mass spectrometry combined with gas chromatography, to deal with the complex mixtures present in sediments. Radiolabelling techniques, widely used in studying biogenic pathways, are now being used to study short-term reaction sequences in sediments.[34] The uniformity of many biochemical processes, in which the same biochemical pathways occur in all living systems, means that many carbon compounds are potential biological markers. Minor variations in the later stages of biosynthetic pathways lead to the possibility of using secondary metabolites for chemotaxonomic purposes. For recent sediments the presence of a particular organism in its environment of deposition may be inferred by identification of a specific marker. When the biolipid has been structurally altered by incorporation in a sediment, the addition of labelled material to the sediment, either *in situ* or under more closely regulated conditions in the laboratory, may be used to follow the rate of conversion and the transformation pathway.

**Hydrocarbons.**—Early studies, in which the hydrocarbon content of sediments from lakes varying in trophic status and geographic location was expressed as a fraction of the total lipoid material, have been reviewed by Swain.[35] The results suggested that oligotrophic lakes contained a lower total lipoid content but a higher proportion of hydrocarbons compared with eutrophic lakes, in which the polar lipoid fraction is dominant. Since g.l.c. techniques

[31] J. R. Vallentyne, *Mitt. int. Verein. theor. angew. Limnol.*, 1969, **17**, 104.
[32] E. S. Deevey in 'Nutrients and Eutrophication', Special Symposium, American Society for Limnology and Oceanography, Lawrence, 1972, pp. 14—20.
[33] W. Pennington, R. S. Cambray, and E. M. Fisher, *Nature*, 1973, **242**, 324.
[34] G. Eglinton, *Pure Appl. Chem.*, 1973, **34**, 611.
[35] F. M. Swain, 'Non-marine Organic Geochemistry', Cambridge Univ. Press, 1970.

were not used, the individual components were not studied. Among hydrocarbons the n-alkanes appear to be useful biological markers because of their relative stability to microbial attack, shown by the persistence in recent sediments of the odd-carbon predominance found in higher plants[36] and the concentrations in sediments which are an order of magnitude greater than that which is found in an average biological lipid.[37] The distribution pattern of n-alkanes isolated from sediments of oligotrophic lakes in N.W. England showed a clear correlation with the type of organic material derived from the drainage basin.[36] The hydrocarbon distribution of the recent sediment of a saline lake showed a bimodal distribution with maxima at $C_{17}$ and $C_{29}$, corresponding with the distributions in the dominant algae and in higher plants, respectively.[38]

Iso-(2-methyl)- and anteiso-(3-methyl)-alkanes have not yet been isolated from recent sediments, although they occur in plant waxes, and the iso- and anteiso-acids of bacterial lipids would be possible precursors.[39] The occurrence of isomeric 6-, 7-, and 8-methylheptadecanes in blue-green algae[40] would appear to represent a useful biological marker, and evidence for the presence of this component in surface sediment samples from a number of lakes was obtained by g.c.—m.s. examination of the branched/cyclic alkane fraction.[41]

The acyclic isoprenoid alkanes pristane (21; R = Me) and phytane (21; R = Et) have been found in sediments of all ages,[39] and the stereochemistry of pristane isolated from an ancient lacustrine sediment indicated an origin from phytol.[42] Pristane and phytane occur in Dead Sea sediment at concentrations of 5 $\mu$g and 3 $\mu$g (100 g dry sediment)$^{-1}$, respectively, and are believed to be derived from abundant halophilic bacteria in the water column of this lake.[43]

The most widespread geolipids derived from terpenoid biolipids are cyclic alkanes, which are stable over long periods of geological time and are good biological markers because of the structural specificity of the polycyclic skeleton. Pentacyclic triterpane hydrocarbons having 27 to 31 carbon atoms have been found in both productive and oligotrophic lakes.[41] A $C_{31}$ homohopane was recently found in an ancient lacustrine sediment, and a derivation from

[36] P. A. Cranwell, *Freshwater Biol.*, 1973, **3**, 259.
[37] W. G. Meinschein in 'Organic Geochemistry. Methods and Results', ed. G. Eglinton and Sister M. T. J. Murphy, Springer-Verlag, Heidelberg, 1969, pp. 330–356.
[38] W. Henderson, W. E. Reed, and G. Steel in 'Advances in Organic Geochemistry 1971', ed. H. R. von Gaertner and H. Wehner, Pergamon Press, Oxford, 1972, pp. 335—352.
[39] J. R. Maxwell, C. T. Pillinger, and G. Eglinton, *Quart. Rev.*, 1971, **25**, 571.
[40] J. Han, E. D. McCarthy, M. Calvin, and M. H. Benn, *J. Chem. Soc.* (*C*), 1968, 2875; J. Han and M. Calvin, *Chem. Comm.*, 1970, 1490.
[41] G. Eglinton, J. R. Maxwell, and R. P. Philp in 'Advances in Organic Geochemistry 1973', in the press.
[42] J. R. Maxwell, R. E. Cox, R. G. Ackman, and S. N. Hooper in 'Advances in Organic Geochemistry 1971', ed. H. R. von Gaertner and H. Wehner, Pergamon Press, Oxford, 1972, pp. 277—291.
[43] A. Nissenbaum, M. J. Baedecker, and I. R. Kaplan, *Geochim. Cosmochim. Acta*, 1972, **36**, 709.

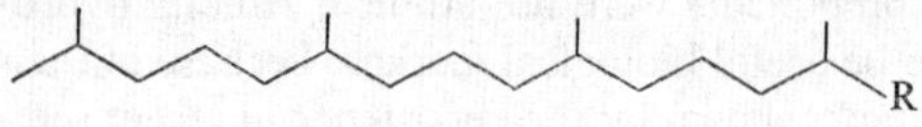


(21)

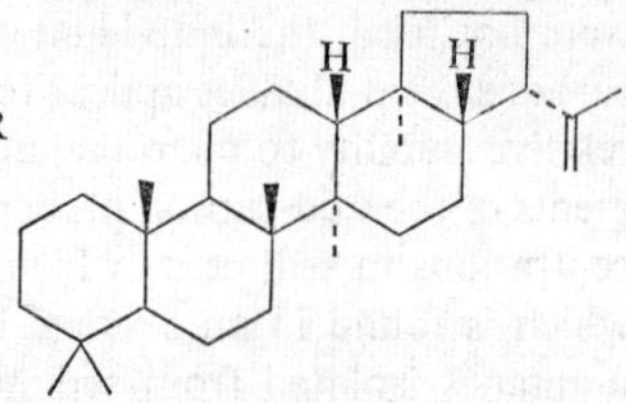


(22)

hop-22(29)-ene (22) by microbial methylation was suggested since pentacyclic triterpanes with more than 30 carbon atoms had not been detected in biological material.[44] This homohopane (A), together with an isomeric pentacyclic triterpane, has been found in the surface sediments of a productive lake, in which (A) was the minor isomer.[45] The other isomer has also been detected in an oligotrophic sediment. Pentacyclic triterpanes found in sediments and petroleum have previously been considered to indicate a higher-plant contribution to the organic material, but the triterpene hop-22(29)-ene together with unidentified $C_{31}$ triterpenes have been isolated from a bacterium,[46] and the presence of hopane-type $C_{35}$ terpenoids in bacteria has been reported.[47]

Aromatic hydrocarbons were detected but not identified in earlier studies. They have been found in rural soils and are not thought to be entirely derived from atmospheric fall-out.[48]

**Fatty Acids.**—n-Alkanoic acids in the range $C_{12}$-$C_{30}$, with $C_{16}$ as a major component, were reported to constitute 0.1% of the dry weight of sediment from a eutrophic lake.[49] Differences in distribution have been observed which may reflect different conditions in the environment; thus in two eutrophic lakes, double maxima at $C_{16}$ and $C_{22}$ are observed, while an oligotrophic lake showed a single maximum at $C_{26}$.[34]

The branched/cyclic acids contain a range of iso- and anteiso-acids, believed to be of microbial origin,[49] and small amounts of isoprenoid acids such as pristanic (21; R = $CO_2H$) and phytanic (21; R = $CH_2CO_2H$).[34] Other components of the branched/cyclic fraction for which a microbial origin has been suggested[50] include 10-methyl branched $C_{16}$ and $C_{18}$ acids and the cyclopropanoid acids dihydrosterculic (23a) and lactobacillic (23b). A capillary g.l.c. trace of a branched/cyclic acid fraction is shown in Figure 1

[44] A. Ensminger, P. Albrecht, G. Ourisson, B. J. Kimble, J. R. Maxwell, and G. Eglinton, *Tetrahedron Letters*, 1972, 3861.
[45] S. J. Gaskell, personal communication.
[46] M. deRosa, A. Gambacorta, L. Minale, and J. D. Bu'Lock, *Chem. Comm.*, 1971, 619; *Phytochemistry*, 1973, **12**, 1117.
[47] H. J. Förster, K. Biemann, W. G. Haigh, N. H. Tattrie, and J. R. Colvin, *Biochem. J.*, 1973, **135**, 133.
[48] R. I. Morrison in 'Organic Geochemistry. Methods and Results', ed. G. Eglinton and Sister M. T. J. Murphy, Springer-Verlag, Heidelberg, 1969, pp. 558—575.
[49] G. Eglinton, D. H. Hunneman, and K. Douraghi-Zadeh, *Tetrahedron*, 1968, **24**, 5929
[50] P. A. Cranwell, *Chem. Geol.*, 1973, **11**, 307.

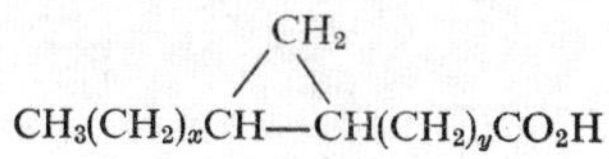


(23a) $x = y = 7$
(23b) $x = 5, y = 9$

and a comparison of saturated monocarboxylic acid distributions in recent and ancient freshwater sediments in Table 1.

The occurrence of considerable quantities of $\alpha,\omega$-dicarboxylic acids in sediments has been reported.[49] The parallel distribution of the chain lengths of these diacids with those of the $\omega$-hydroxy-acids was believed to indicate a derivation from the latter. A series of $\alpha$- and $\beta$-hydroxy-acids was also found in this sediment, having a chain-length distribution parallel with that of the fatty acids, suggesting that they were intermediates in the degradation of fatty acids.

The presence has been reported of both free and esterified fatty acids in sediments, with a considerable difference in composition and chain-length distribution between the two fractions.[43]

The stereochemistries of the isoprenoid acids from an ancient sediment are compatible with a phytol origin,[39] but stereochemical studies on samples from recent sediments have not been reported. However, the conversion of $^{14}$C-labelled phytol into phytanic and pristanic acids in a sediment has recently been reported.[51]

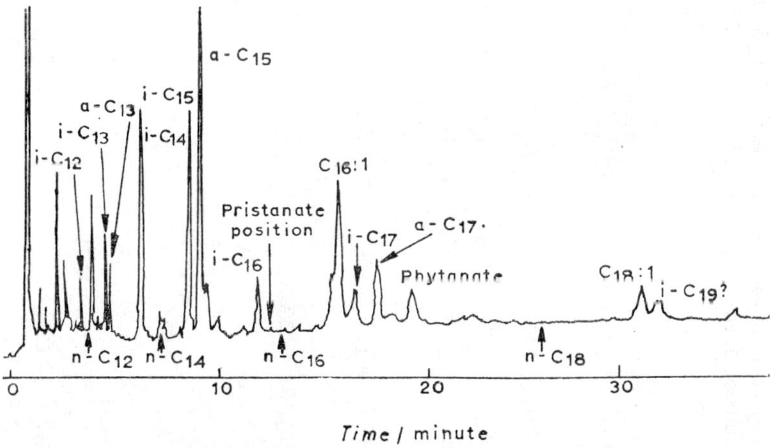


**Figure 1** *Gas chromatogram of urea non-adduct fraction of total fatty acids from the sediment of Esthwaite Water. Some straight-chain acids are still present. Conditions: acids as methyl esters,* 50 m *capillary column coated with butanediol succinate, isothermal at* 170 °C, 40 p.s.i. Reproduced by permission from *Pure Appl. Chem*, 1973, **34,** 611)

[51] P. W. Brooks and J. R. Maxwell in 'Advances in Organic Geochemistry 1973', in the press.

**Table 1** *Comparison of saturated fatty-acid contributions in Recent and Ancient freshwater sediments*
(Reproduced by permission from *Pure Appl. Chem.*, 1973 **34**, 611)

| | | Recent | | | | | | Ancient | |
|---|---|---|---|---|---|---|---|---|---|
| | | *Blelham* | | *Esthwaite* | | *Ennerdale* | | *Green River Shale* | |
| | *Acid type* | *Main*[a] | *% of total* | *Main*[a] | *% of total* | *Main*[a] | *% of total* | *Main*[a] | *% of total* |
| Normal | $CH_3(CH_2)_nCO_2H$ | **16**, *18*, *20*, **22**, 24 | 70 | **16**, *18*, *20*, **22**, 24 | 86 | 24, **26**, 28 | 93 | 16, 18, *28*, **30** | 80 |
| Iso | $CH_3CH(CH_3)(CH_2)_nCO_2H$ | *14*, **15**, 16, 17 | 8 | *14*, *15*, 16, 17 | 6 | **15**, 16, *17* | 7 (Iso to Isoprenoid) | 14, 15, 16, 18 | 8 (Iso and Anteiso) |
| Anteiso | $CH_3CH_2\overset{*}{C}H(CH_3)(CH_2)_nCO_2H$ | **15**, *17* | 12 | **15**, *17* | 6 | *15*, **17**, 19 | | 15 | |
| Branched | $CH_3(CH_2)_n\overset{*}{C}H(CH_3)(CH_2)_mCO_2H$ | **17**, *19* | 2 | — | — | 17, **19** | | — | — |
| Cyclic | $CH_3(CH_2)_n\overset{*}{C}H$–$\overset{*}{C}H(CH_2)_mCO_2H$ (bridged by $CH_2$) | 17, 19 | 2 | — | — | 17, *19* | | — | — |
| Isoprenoid | $H(CH_2$–$^*CH(CH_3)$–$CH_2$–$CH_2)_n$–X; X = $CH_2$–$^*CH(CH_3)$–$CH_2$–$CO_2H$ | 20 | 4 | 20 | 1 | 19, **20** | | 19, 20 | 12 |

[a] Relative abundance **x** > *x* > x.

Unsaturated acids which are typical of the plankton in the overlying water have been reported by Poltz,[52] who estimated that only 1—2.5% of the total lipids produced within the lake, 0.1—0.5% of the total fatty acids, and less than 0.1% of the triglycerides are found in the sediments. The rapid decomposition of unsaturated acids in the upper sediment of Lake Windermere was followed by injection of [*U*-$^{14}$C]oleic acid into a sealed sediment core. After three weeks, only 10% of the labelled oleic acid remained, but labelled $C_{14}$, $C_{16}$, and $C_{18}$ n-alkanoic acids were also identified and accounted for 36% of the initial radioactivity.[53] It was suggested, from the ratio of saturated:unsaturated acids of oxidizing and reducing sediments in the Dead Sea, that the unsaturated acids are altered at the same relative rate in both types of environment.[43]

**Alcohols and Sterols.**—The isolation of n-alcohols from recent marine and ancient freshwater sediments has been reviewed,[39] while a series of n-alcohols in the range $C_{16}$—$C_{30}$, showing strong even/odd predominance parallel to the distribution in waxes of higher plants, has been identified in sediments from several lakes.[54] Dihydrophytol and phytol (24) have been reported in deep-water reducing sediments of the Dead Sea in concentrations up to 81 and 28 $\mu$g (100 g dry sediment)$^{-1}$, respectively.[43] Phytol, the diterpenoid side-chain of chlorophyll *a* (25), is thought to be the major precursor of the

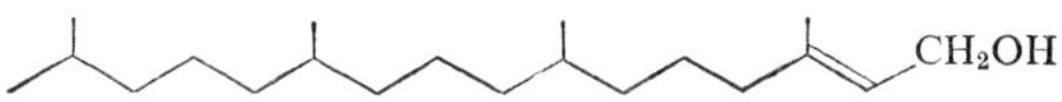


(24)

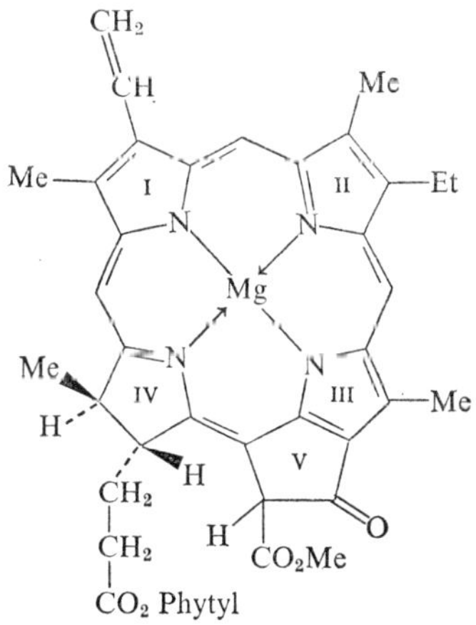


(25)

[52] J. Poltz, *Arch. Hydrobiol., Suppl. 40*, 1972, 315.
[53] M. M. Rhead, G. Eglinton, G. H. Draffan, G. Ware, P. England, and F. J. H. Mackereth, 1971, unpublished observations.
[54] P. A. Cranwell, *Ann. Rep. Freshwater Biological Association*, 1971; 1973.

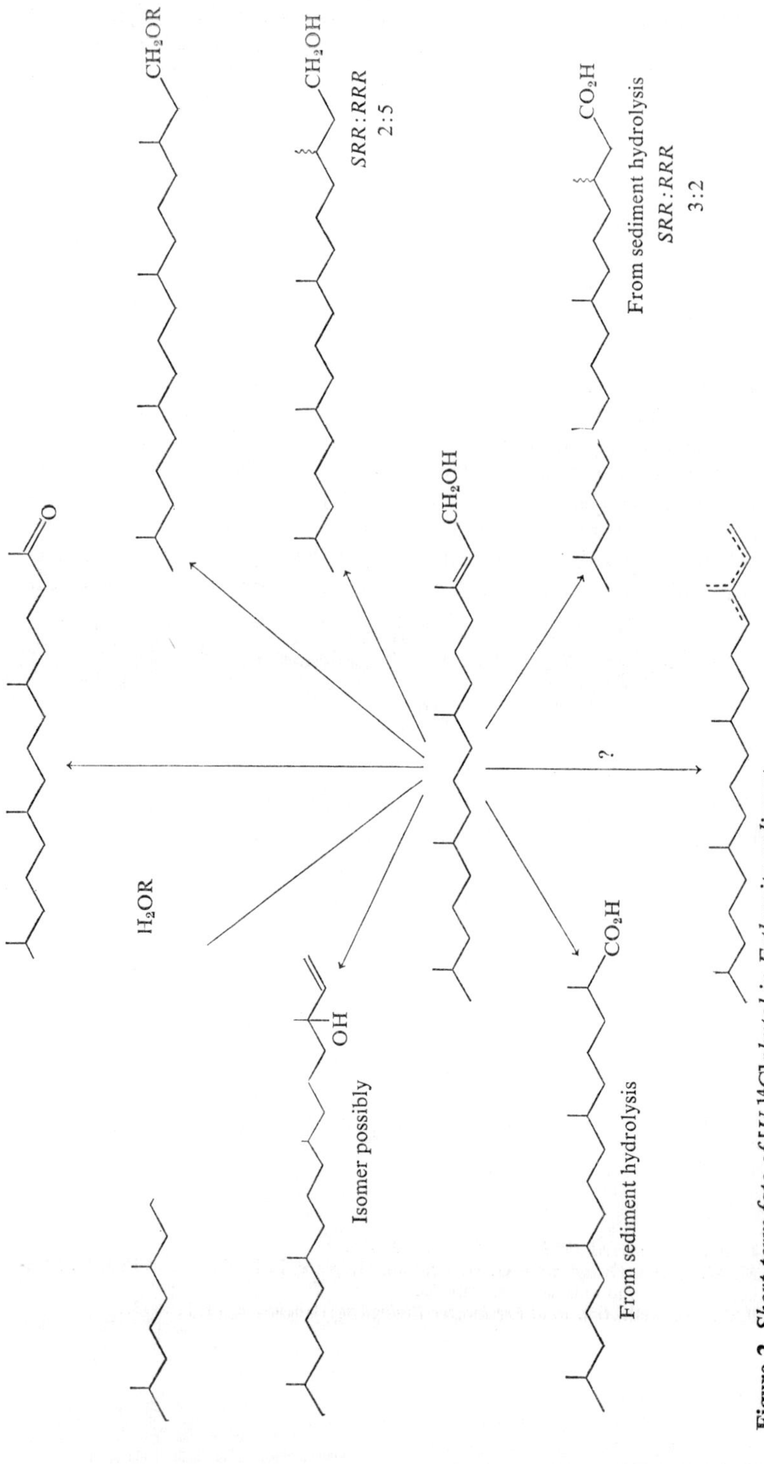


**Figure 2** *Short-term fate of* [*U*-[14]C]*phytol in Esthwaite sediment*
(Reproduced by permission from 'Advances in Organic Geochemistry 1973', in the press)

acyclic isoprenoids with carbon number of 20 or less which occur in the geosphere.[39] The early stages in the diagenesis of phytol have been examined by incubation of [*U*-$^{14}$C]phytol in surface sediment samples taken from a productive lake.[51] The results are summarized in Figure 2 and Table 2.

**Table 2** *Radiolabelled products of incubation of* [*U*-$^{14}$C]*phytol with Esthwaite sediment*

| *Fraction* | *Major labelled compound present* | *Concentration in unlabelled sediment* |
| --- | --- | --- |
| Alcohols | Phytol | *ca.* 5 p.p.m. |
| | Dihydrophytol | *ca.* 0.03 p.p.m. |
| | Phytol isomer | |
| Esters | Phytol[a] | *ca.* 0.13 p.p.m.[b] |
| Ketones | 6,10,14-Trimethyl-pentadecan-2-one | *ca.* 2 p.p.m. |
| Dienes | Phytadienes | not detected |
| Acids[c] | Phytanic | *ca.* 0.01 p.p.m. |
| | Pristanic | *ca.* 0.02 p.p.m. |

[a] After hydrolysis of esters in neutral extract;
[b] Mass calculated as free alcohol;
[c] In acid hydrolysis of the sediment.

The presence of sterols in freshwater sediments at concentrations up to 300 parts of cholesterol carbon per million parts of organic carbon was reported in 1964.[39] 5α- and 5β-Stanols corresponding to the Δ$^5$-sterols (26;

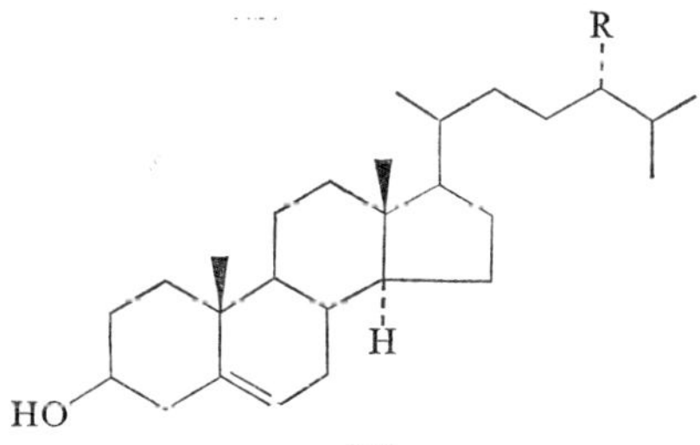


(26)

R = H, Me, or Et), which are common in the biosphere, have been identified both in an Eocene sediment,[55] where they were considered to be intermediates between Δ$^5$-sterols and 5α- and 5β-steranes, and also in a Recent sediment from Mono Lake.[38] The ratio of stanol to sterol concentration increased with depth down a 200 m core from Lake Biwa.[56] A more accurate indication of the time-scale involved in these changes has been obtained by the identification of 5α- and 5β-stanols in the uppermost sediment (less than 10 years old) from a productive lake.[57] Radiolabelling experiments in which [4-$^{14}$C]cholesterol

[55] G. Steel and W. Henderson, *Nature*, 1972, **238**, 148.
[56] K. Ogura and T. Hanya, *Proc. Japan Acad.*, 1973, **49**, 201.
[57] S. J. Gaskell and G. Eglinton in 'Advances in Organic Geochemistry 1973', in the press.

was incubated for three months in a sediment core (from this lake) in the laboratory showed a lower conversion (0.2%) than that expected on the basis of sterol analyses, possibly indicating inadequate simulation of sedimentary conditions.

**Ketones.**—A homologous series of methyl ketones with 19–33 carbon atoms and strong odd/even predominance, as found in soils,[48] has been obtained from several sources.[54] It has been suggested that they are formed by microbial attack on alkanes.[48] 6,10,14-Trimethylpentadecan-2-one has been isolated from the surface sediment of a productive lake. It has been shown that the ketone is derived from phytol by addition of [$^{14}$C]phytol to a sediment and isolation of the labelled ketone, after a suitable incubation period (see Figure 2).[51]

**Carbohydrates.**—The carbohydrates of aquatic source organisms have been reviewed by Swain,[35] who suggested that in a productive lake between 1 and 20% of the quantity of sugars in aquatic plants was preserved in the sediments. Very little free sugar was present in two lakes (described in ref. 35), but hydrolysable sugars which could be identified reached 28 mg (g dry sediment)$^{-1}$ in one sample. In more recent studies of the free and hydrolysable sugars of sediments from a series of lakes differing in trophic status a correlation of high sugar content with high productivity was observed.[58] Glucose (2) [48 mg (l fresh sediment)$^{-1}$], maltose (27), and sucrose (19) were the dominant free sugars of a productive lake, and glucose was dominant in the seston. After hydrolysis, galactose (1), glucose (2), mannose (3), arabinose (4), xylose (5) [up to 600 mg (l fresh sediment)$^{-1}$], fucose (7), ribose (28), rhamnose (6),

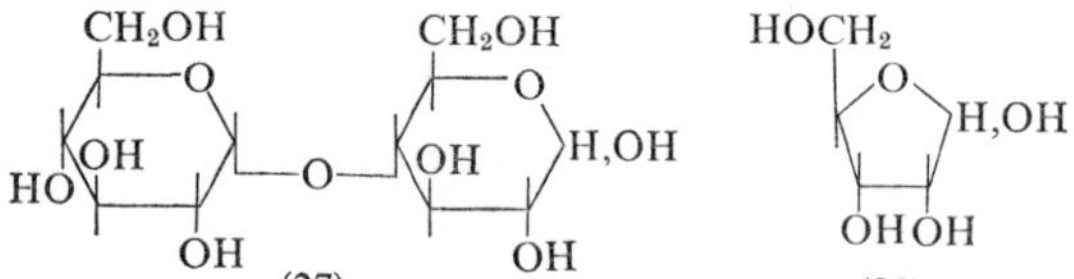

(27) (28)

deoxyribose (8), and fructose (20) were found. In oligotrophic lakes the sugar pattern of the seston changed during sedimentation, so that galactose [up to 470 mg (l sediment)$^{-1}$] and not glucose was the dominant hydrolysable sugar in the surface sediment. Only traces of free sugars were found in oligotrophic lake sediments.

An improved method for determination of total carbohydrate in sediments by the phenol–sulphuric acid method has recently been described.[59]

**Amino-acids.**—The hydrolysable amino-acid content of a series of lake sediments reported by Swain[35] showed a broad correlation with trophic status, ranging from 0.3 to 41 parts per ten thousand. No correlation was

[58] S. Fleischer, *Arch. Hydrobiol.*, 1972, **70**, 392.
[59] S. M. Gerchakov and P. G. Hatcher, *Limnol. Oceanogr.*, 1972, **17**, 938; D. Liu, P. T. S. Wong, and B. J. Dutka, *Water Res.*, 1973, **7**, 741.

observed between the source organisms and sedimentary amino-acid distribution, because of the unequal sensitivity to decomposition. Those sediments in which organic material was accumulated under reducing conditions, thus protecting sensitive amino-acids from decay, showed the greatest diversity of amino-acids in the hydrolysate. Thus, in oxidizing and reducing sediments of the Dead Sea, the amino-acids constituted 1.5—3.2% and 7.6—11.7% of organic carbon in the respective environments.[43] Differences in distribution were reported, with the aromatic amino-acids tyrosine (29) and phenylalanine (30) representing a much greater fraction of the total in the oxidizing sediment, and the acidic amino-acids aspartic (9; R = $HO_2CCH_2$) and glutamic

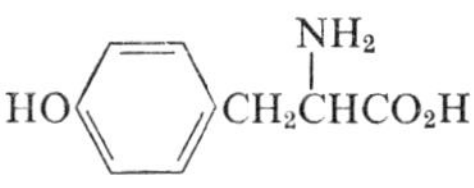

(29)

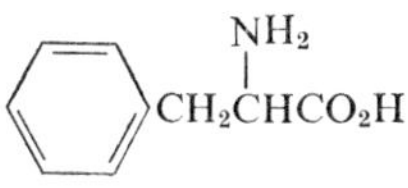

(30)

(9; R = $HO_2CCH_2CH_2$) being much more abundant in the reducing sediments, though a difference in origin of the amino-acids could not be ruled out as an explanation.

Other forms of nitrogen have been reported. Studies on the sediments of Lake Ontario[60] have shown that 90% of N in the surface mud is organic nitrogen [2400 $\mu$g (g dry sediment)$^{-1}$], of which 28—46% was amino-acid nitrogen, 4.7% was hexosamine nitrogen, and 21—31% was unidentified hydrolysable nitrogen, but free amino-acids and soluble combined amino-acids and amino-sugars accounted for less than 0.25% of the total nitrogen. It was estimated that at least 20% of organic nitrogen input to the sediment was regenerated to the lake from the top 6 cm. The organic C:N ratio was approximately constant in three Great Lakes examined. The decrease in the content of hexosamine (glucosamine and galactosamine) with an increase in fertility of a series of lakes has been interpreted in terms of a greater net nitrogen turnover by micro-organisms in oligotrophic than in eutrophic lake sediments.[61]

**Pigments.**—Photosynthetic pigments from bacteria, algae, and higher plants have been found in freshwater sediments.[62,63] The two major groups are the carotenoids and chlorinoid pigments, both of which have been considered as useful indicators in palaeolimnology. Thus the presence in lake sediments of myxoxanthophyll (31) and oscillaxanthin (32), which are restricted in distribution to blue-green algae, has been used as a selective indicator of

[60] A. L. W. Kemp and A. Mudrochova, *Limnol. Oceanogr.*, 1972, **17**, 855; *Geochim. Cosmochim. Acta*, 1973, **37**, 2191.
[61] D. R. Keeney, J. G. Konrad, and G. Chesters, *J. Water Pollution Control Fed.*, 1970, **42**, 411.
[62] J. R. Vallentyne in 'Comparative Biochemistry of Photoreactive Systems', ed. M. B. Allen, Academic Press, New York, 1960, pp. 83—105.
[63] S. R. Brown, *Mitt. int. Verein. theor. angew. Limnol.*, 1969, **17**, 95.

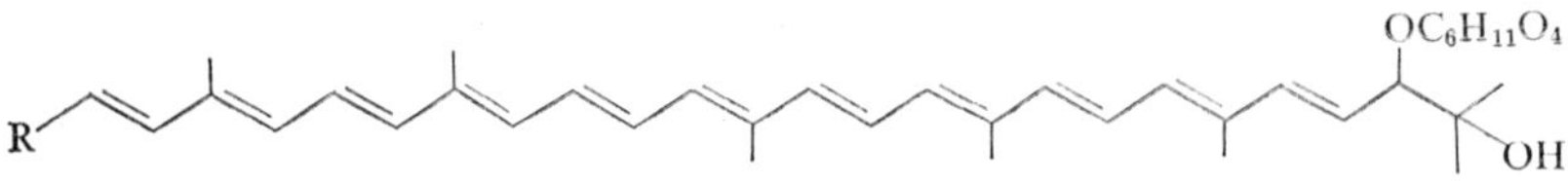

(31) R = 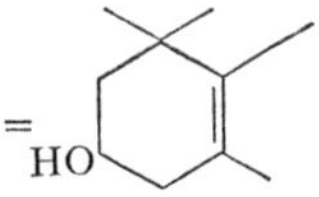 (32) R = 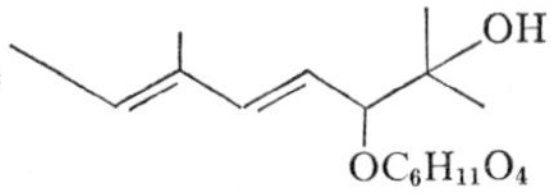

former populations of these algae.[63] Brown used a sedimentary pigment assemblage consisting of spheroidenone (33), spheroidene, and 2-keto-spirilloxanthin (34) to deduce the presence of photosynthetic bacteria of the genus *Rhodopseudomonas* at a period in the development of a lake.[63]

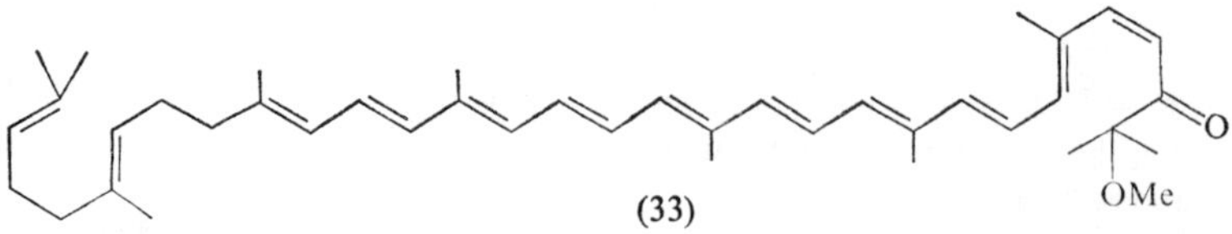

(33)

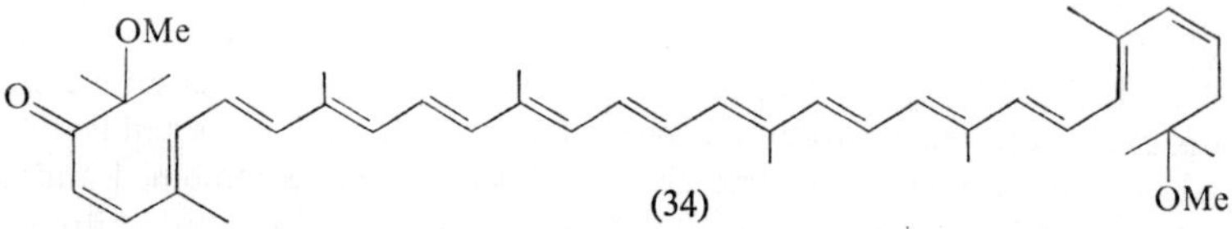

(34)

Evidence from several sources (reviewed by Swain)[35] suggested that chlorophyll degradation products preserved in sediments provided a more sensitive index of lake productivity than the organic carbon content. Thus in surface sediments from a series of lakes of increasing trophic status in which the total carbon increased three-fold (7—19% of dry wt.), chlorophyll derivatives increased thirty-fold [0.2—6.9 units (g sediment)$^{-1}$].[63] Recent studies have attempted to use pigments to elucidate further the sources of organic matter in lake sediments. Although chlorophylls have a limited value in identifying sedimentary source materials, chlorophyll *c* can be used to distinguish certain types of freshwater flora, while bacteriochlorophylls and chlorobium chlorophylls characterize certain groups of bacteria.[63] Reversed-phase t.l.c. has recently been used to resolve the mixture of chlorophylls and their degradation products which occur in lake sediments,[64] and quantitative results have been reported for chlorophylls *a* and *b*, pheophytins *a* and *b*, pheophorbides *a* and *b*, and allomerized *a* and *b* chlorin pigments in sediments of the Great Lakes.[65] As a means of assessing the relative contribution of allochthonous and autochthonous organic matter to lake sediments, Sanger and Gorham have examined the pigment diversity in sediments of productive and unproductive lakes, living and decomposed terrestrial vegetation,

[64] R. J. Daley, C. B. J. Gray, and S. R. Brown, *J. Chromatog.*, 1973, **76**, 175.
[65] C. B. J. Gray and A. L. W. Kemp, *Proc. 13th Conf. Great Lakes Res.*, 1970, 242.

aquatic macrophytes, and algae.[66] Productive lakes contained a greater number of pigments than unproductive lakes (averages from several sources were 38 and 30 pigments, respectively), and a greater concentration of chlorophylls was also noted in the former [16 arbitrary units (g organic matter)$^{-1}$] compared with oligotrophic lakes, in which values near to 1 unit (g organic matter)$^{-1}$ were observed. To account for these observations it was suggested that the more sensitive pigments were destroyed under the aerobic conditions which prevailed all the year at the sediment surface in oligotrophic lakes. Evidence in support of this hypothesis was obtained from a sediment taken from a permanently anaerobic environment. This sample showed an even greater pigment diversity and higher chlorophyll content [50 units (g organic matter)$^{-1}$]. Comparison of the pigment diversity in possible source materials indicated that algal decomposition was mainly responsible for the diversity. However, the only components identified during these studies were lutein (35) and $\beta$-carotene (36), both of these derived from terrestrial detritus.

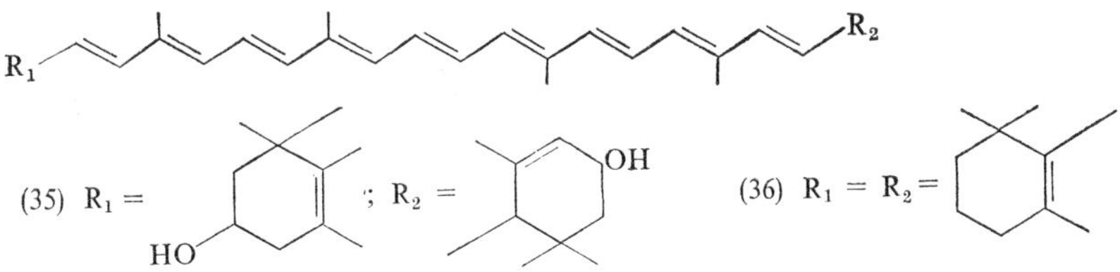


Chlorophyll diagenesis in fresh waters has been examined and two unrelated processes have been observed.[67] These were chlorophyll destruction (rupture of tetrapyrrole ring) and derivative formation (partial degradation to coloured identifiable derivatives). The principal diagenetic mechanisms of phytoplankton chlorophylls are believed to be post-lytic photochemical oxidation and herbivore predation. It was suggested that the pheophorbide content of sediments was a measure of herbivore grazing pressure and that the allomerized phorbin content was a measure of anoxia in the water column and, therefore, an indirect measure of lake trophic development.

**Sedimentary Humus**—The humic substances of aquatic sediments have received much less attention than those of soils, but i.r. spectroscopy was used to distinguish the humic acid fraction of a sediment thought to be of autochthonous origin from the humic acid of the soil within the watershed.[68] Ishiwatari[69] has studied a number of sedimentary humic acid samples by spectroscopic techniques and by the application of degradation methods used on soil humic materials. He concluded that sedimentary humus was derived from both terrestrial and lacustrine sources.

[66] J. E. Sanger and E. Gorham, *Limnol. Oceanogr.*, 1970, **15**, 59; 1972, **17**, 618, 840
[67] R. J. Daley and S. R. Brown, *Arch. Hydrobiol.*, 1973, **72**, 277; R. J. Daley, *ibid.*, p. 409.
[68] A. Otsuki and T. Hanya, *Geochim. Cosmochim. Acta*, 1967, **31**, 1505.
[69] R. Ishiwatari, *Chem. Geol.*, 1973, **12**, 113.

## 5 Organophosphorus Compounds in Water and Sediments

In spite of the importance of phosphorus in the eutrophication of lakes and the knowledge that organically bound phosphorus is present in significant amounts (up to 50 $\mu$g $PO_4$ $l^{-1}$) in natural waters, the nature and identity of the compounds have received little attention.[70,71] It was estimated that less than 5% of dissolved organic phosphorus in a eutrophic lake was nucleic acid or phospholipid,[71] and recent molecular-sieve studies on natural lake water samples showed that up to 20% of organic phosphorus has a molecular weight exceeding 50 000, which included some DNA or DNA fragments.[72] Methods for determination of phosphate and total phosphorus in waters have been reviewed[71] and methods of estimating the total organic phosphorus in lake sediments were compared by Sommers *et al.*,[73] who found 23—147 mg P (100 g dry sediment)$^{-1}$, in which the organic P was 8—63% of the total phosphorus.

Exchange of phosphorus between sediment and water is the most important aspect of the chemistry of the element. In most lakes there is a net removal of phosphorus into the sediment, mainly by adsorption on ferric hydroxides and oxides, but when the redox potential decreases, solution of ferrous iron occurs and phosphorus is released. The reduction in redox potential that causes phosphorus release corresponds to a dissolved oxygen concentration of 2 mg $l^{-1}$, so that phosphorus release can occur during deoxygenation of the hypolimnetic waters of a stratified lake. Solubilization of inorganic phosphate by bacteria isolated from a lake sediment was shown to be associated with carbohydrate metabolism and aerobic growth, and was attributed to the chelating ability of the released organic acids.[74]

## 6 Chemical Pollution of the Aquatic Environment

Pollution may be equated with an unfavourable alteration of the environment. Three categories of pollution have been distinguished:

(*a*) Pollution by toxic agents that directly, or after incorporation in the food chain, represent a hazard to man.

(*b*) Pollution resulting from a departure of a balance between photosynthesis and respiration; this type of disturbance is usually caused by enrichment with organic wastes or algal nutrients and has been termed eutrophication.[75]

[70] 'Phosphorus in Fresh Water and the Marine Environment', ed. S. H. Jenkins, *Water Research*, 1973, **7**, pp. 1—342.
[71] J. R. Kramer, S. E. Herbes, and H. E. Allen in 'Nutrients in Natural Waters', ed. H. E. Allen and J. R. Kramer, Wiley, New York, 1972, pp. 51—100.
[72] R. A. Minear, *Environ. Sci. Technol.*, 1972, **6**, 431.
[73] L. E. Sommers, R. F. Harris, J. D. H. Williams, D. E. Armstrong, and J. K. Syers, *Limnol. Oceanogr.*, 1970, **15**, 301.
[74] M. J. Harrison, R. E. Pacha, and R. Y. Morita, *Limnol. Oceanogr.*, 1972, **17**, 50.
[75] J. W. G. Lund, *Proc. Roy. Soc.*, 1972, **B180**, 371.

(*c*) Pollution as an impairment of the diverse ecosystems constituting the biosphere; simplification and decrease in stability of an ecosystem reduce its survival value.

The sources of organic chemicals mainly responsible for these types of pollution are, for type (*a*), persistent pesticides and organo-mercury compounds, and for type (*b*) sewage effluents and drainage containing agricultural fertilizers. Pollution of type (*c*) arises from the influence of man on the ecological succession through agricultural practices and other activities which reverse natural energy flows in an unpolluted system.[76]

A review[77] of organic chemical pollution of freshwater up to mid 1970 noted 66 compounds which had been identified in freshwater, though one of these, fulvic acid, had been incorrectly assigned the structure of a microbial product of the same name. Industrial sources were responsible for the largest number and variety of structural types, while reported agricultural sources of pollutants were all pesticides.

Techniques for extraction, concentration, and purification of organic pollutants prior to analysis have been reviewed.[7,78] Identification of organochlorine pesticides has been achieved by a combination of multiple-column g.l.c. with electron-capture detection and loose-layer t.l.c.[79] Background levels in non-polluted waters of the United Kingdom were: α-BHC (benzene hexachloride) 10—20 ng $l^{-1}$; γ-BHC 10—60 ng $l^{-1}$; Dieldrin (37) 0—20 ng $l^{-1}$; *pp*′-DDT (38; R = $CCl_3$) $<5$ ng $l^{-1}$ (except during orchard spraying). The detection limits ranged from 0.3 ng $l^{-1}$ for α-BHC to 5 ng $l^{-1}$ for *pp*′-DDT. A method for the chromatographic separation of 50 pesticides isolated from water samples into 4 groups has been described.[80] Triazine herbicides [*e.g.* simazine (40)], used in aquatic weed control, have been analysed by g.l.c. using an alkali flame detector with a sensitivity of 0.001 p.p.m.[81] G.c.–m.s. has been used in the separation and identification of compounds extracted from polluted river waters[82,83] and potable water supplies,[84] while both field-desorption and electron-impact mass spectrometry have been used to identify pesticides and their metabolites.[85]

76 W. Stumm and E. Stumm-Zollinger in 'Nonequilibrium Systems in Natural Water Chemistry', Advances in Chemistry Series, No. 106, Amer. Chem. Soc., Washington, 1971, pp. 1—29.
77 Water Quality Criteria Data Book, Vol. 1, Water Pollution Control Research Series, 18010DPV, Environmental Protection Agency, Washington, 1970.
78 S. D. Faust and I. H. Suffet in 'Microorganic Matter in Water', American Society for Testing and Materials STP 448, Philadelphia, 1969, pp. 24—64; W. Leithe, 'Analysis of Organic Water Pollutants', Ann Arbor Science Publishers, Michigan, 1973, 213 pp.
79 B. T. Croll, *Water Treatment and Examination*, 1969, **18**, 255.
80 V. Leoni, *J. Chromatog.*, 1971, **62**, 63.
81 C. E. McKone, T. H. Byast, and R. J. Hance, *Analyst*, 1972, **97**, 653.
82 R. A. Hites and K. Biemann, *Science*, 1972, **178**, 158.
83 R. D. Kleopfer and B. J. Fairless, *Environ. Sci. Technol.*, 1972, **6**, 1036.
84 K. Grob, *J. Chromatog.*, 1973, **84**, 255.
85 H. R. Schulten, H. Prinz, H. D. Beckey, W. Tomberg, W. Klein, and F. Korte, *Chemosphere*, 1973, **2**, 23; J. N. Damico in 'Biochemical Applications of Mass Spectrometry', ed. G. R. Waller, Wiley, New York, 1972, pp. 623—653.

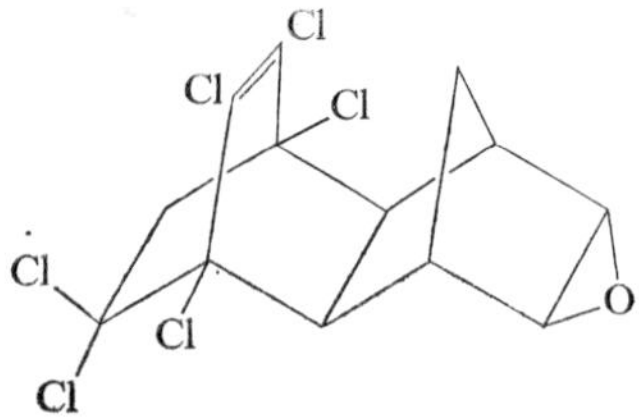

**(37)**

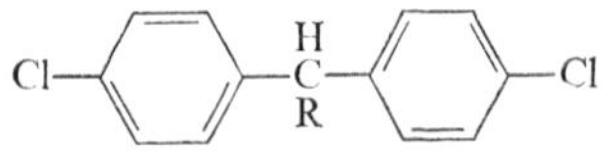

(38)

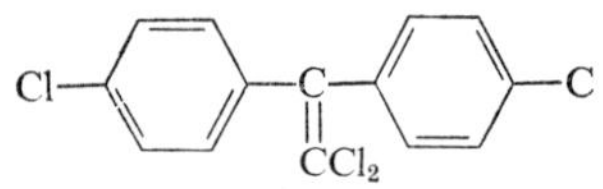

(39)

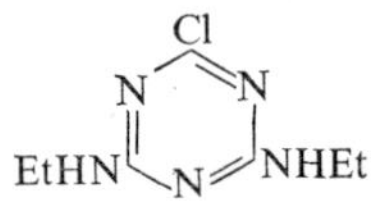

(40)

**Pesticides.**—The persistence of pesticides within a watershed, the mechanism of pesticide transport into the aquatic system, and features within the aquatic system which alter the distribution of the adsorbed pesticide between water and associated sediment within the lake have been reviewed recently.[86]

The accumulation of chlorine-containing pesticides has been reported for all levels of the food chain in the aquatic environment. In laboratory streams containing 0.05—7 $\mu g\,l^{-1}$ of Dieldrin (37), benthic algae were found to concentrate the pesticide by factors up to 30 000 over a period of four months,[87] while *Daphnia* concentrated DDT, from aqueous suspensions containing down to 8 $\mu g\,l^{-1}$, by a factor of 16 000—23 000 within 24 h.[88] An increase in the amount of DDT and PCB's (polychlorinated biphenyls) in trout has been reported with increasing maturity.[89]

The power of sediments and colloidal organic matter to absorb pesticides has been investigated. After application of Toxaphene (a mixture of chlorinated terpenes) to a lake (3 times at 0.15 $mg\,l^{-1}$), the concentration in the sediments increased for 3 months, reaching 15.8 $\mu g\,g^{-1}$ in the upper sediment, then decreasing by a factor of 2 every 4 months. Transport down the sediment profile as far as the 20 cm horizon was also shown to occur.[90] The observed 15 800-fold concentration of $^{14}C$-labelled DDT on the colloidal organic matter of a natural brown water containing 0.17 $\mu g\,l^{-1}$ of the labelled pesticide suggests a mechanism by which pesticides can be transported in moving waters, or can be added to the sediment when the colloids precipitate.[91] The use of Paraquat (41) for aquatic weed control has led to a study of its residence time and movement within water bodies. After application to a reservoir at 1 $mg\,l^{-1}$, no Paraquat could be detected in the water after 2 weeks, but could

86 H. B. Pionke and G. Chesters, *J. Environ. Quality*, 1973, **2**, 29.
87 F. L. Rose and C. D. McIntire, *Hydrobiologia*, 1970, **35**, 481.
88 D. G. Crosby and R. K. Tucker, *Environ. Sci. Technol.*, 1971, **5**, 714.
89 C. A. Bache, J. W. Serum, W. D. Youngs, and D. J. Lisk, *Science*, 1972, **177**, 1191.
90 G. D. Veith and G. F. Lee, *Environ. Sci. Technol.*, 1971, **5**, 230.
91 M. A. Poirrier, B. R. Bordelon, and J. L. Laseter, *Environ. Sci. Technol.*, 1972, **6**, 1033.

be identified in the top 6 cm of sediment.[92] However, after 8 months only 64% of the applied Paraquat could be detected in the sediment.

The presence of the organophosphorus pesticides Malathion (42) and Diazinon (43) has been detected in river waters at concentrations up to 10 ng $l^{-1}$ and 10—30 ng $l^{-1}$, respectively.[93]

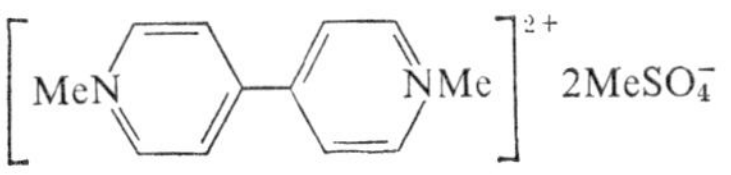

(41)

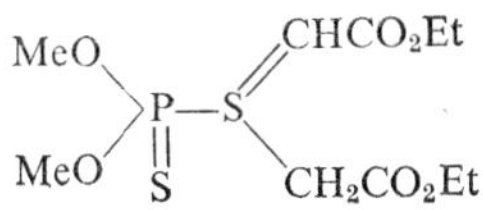

(42)

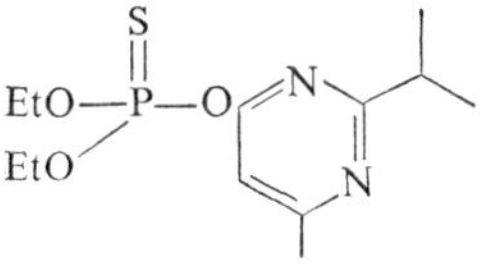

(43)

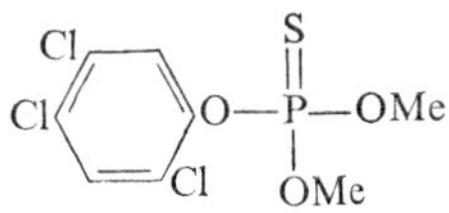

(44)

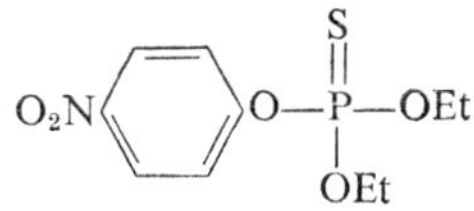

(45)

Techniques for recovery, separation, and isolation of organophosphates from water have been devised which take into account their reactivity, so that parent, oxo-derivative formed by desulphuration, and hydrolysis product may all be recovered.[78] Leoni detected Ronnel (44), Parathion (45), Malathion, and Diazinon in polluted waters using a separation procedure mentioned above.[80]

*Stability and Fate of Pesticides.* Studies on the persistence of a range of organochlorine, organophosphorus, and carbamate pesticides in river water, using initial concentrations of 10 $\mu g\ l^{-1}$, showed that, with one exception, all of the carbamates and organophosphorus compounds were degraded, but over half of the organochlorine compounds were unchanged.[94]

Photodecomposition of pesticides has been studied under laboratory conditions.[95] Almost all of the sun's emitted radiation of wavelength shorter than 285 nm is absorbed by ozone in the upper atmosphere though, because the absorption coefficient of ozone decreases sharply at 220 nm, some energy in the 200—220 nm region reaches the earth. Only those compounds which absorb light energy in these available regions undergo a photoreaction. Types of photochemical reaction which have been observed are photoisomerization, especially in the case of chlorinated cyclodiene insecticides, and aryl–halogen bond fission in halogenated aromatic pesticides, to give phenols

92 M. P. Brooker and R. W. Edwards, *Freshwater Biol.*, 1973, **3**, 157.
93 G. F. Lowden, C. L. Saunders, and R. W. Edwards, *Water Treatment and Examination*, 1969, **18**, 275.
94 J. W. Eichelberger and J. J. Lichtenberg, *Environ. Sci. Technol.*, 1971, **5**, 541.
95 J. D. Rosen in 'Organic Compounds in Aquatic Environments', ed. S. D. Faust and J. V. Hunter, Dekker, New York, 1971, pp. 425—438.

as the major products. Under environmental conditions pesticides may be in contact with materials which act as photosensitizers. Energy may thus be transferred to the pesticide, causing photolysis of pesticides which do not absorb light.

The photodegradation of certain carbamate insecticides in aqueous solution using u.v. light (254 nm) was investigated and it was shown that further photolysis of the initial decomposition products occurred.[96] Under these conditions the dialkyl carbamates Pyrolan (46) and Dimetilan (47) were decomposed more rapidly than the monoalkyl carbamates Baygon (48) and Sevin (49). Under alkaline hydrolysis conditions the order of stability was reversed,

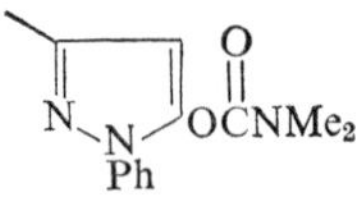
(46)

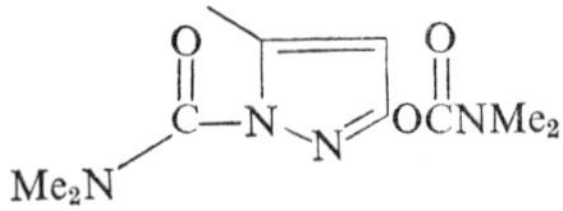
(47)

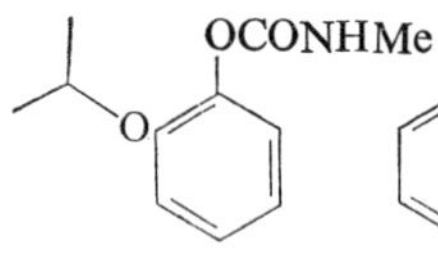
(48)

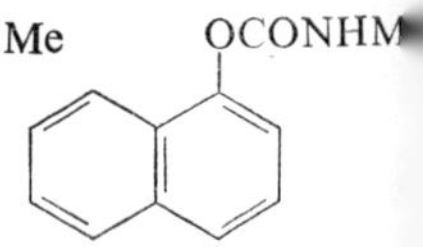
(49)

since the monoalkyl carbamates were hydrolysed at pH values higher than 7 while the dialkyl carbamates were stable towards hydrolysis in the pH range 4—10.

The biodegradability and fate of DDT and its metabolites in the aquatic environment were studied[97] using a laboratory ecosystem with a terrestrial–aquatic interface and a seven-element food chain. $^{14}C$-Labelled DDT was accumulated in the food chain as DDT (38; R = $CCl_3$), DDE (39), and DDD (38; R = $CHCl_2$), with $10^4$—$10^5$ fold concentration. [$^{14}C$]DDE was concentrated 3—5 × $10^4$ fold and was less biodegradable than DDT since it was stored in the lipids of the food chain largely unchanged, only 34% of the label being present as polar lipids in the water, compared with 80% of the $^{14}C$ when [$^{14}C$]DDT was used. A similar study, using an ecosystem containing both producer and consumer organisms, showed that DDT and DDE were taken up mainly by algae, which conveyed the pesticide to the first consumer organism. The producer organisms were unable to decompose either DDT or DDE, but consumer organisms converted DDT into DDE and DDD.[98]

Some pesticides are resistant to conventional water and waste-water treatments and may resist biologically mediated degradation in aquatic environments, where they persist for a considerable time. Several methods of removing pesticides from aqueous solution have been examined. Kinetic studies of the adsorption of pesticides by active carbon showed that a similar rate of

[96] O. M. Aly and M. A. El-Dib, *Water Res.*, 1971, **5**, 1191; also in 'Organic Compounds in Aquatic Environments', ed. S. D. Faust and J. V. Hunter, Dekker, New York, 1971, pp. 469—493.

[97] R. L. Metcalfe, G. K. Sangha, and I. P. Kapoor, *Environ. Sci. Technol.*, 1971, **5**, 709.

[98] A. Södergren, *Oikos*, 1973, **24**, 30.

removal was observed for a broad spectrum of compounds.[99] Chemical oxidants have been used to reduce the concentration of contaminants in treatment plants. The feasibility of a chemical oxidative treatment may be determined from free-energy of formation data for the reactants and products if some assumptions are made about the structure of the products, but the overall rate of the process is dependent on kinetic factors. Knowledge of these kinetic and thermodynamic factors was used to determine the most suitable reagent and optimum conditions for oxidation of dipyridylium quaternary herbicides in water.[100]

**Industrial and Domestic Pollutants and Sewage Effluents.**—The review[77] mentioned above, of organic pollution of freshwater, also contains details of the suspected source of some pollutants. Particular interest has been focussed on rivers and the distance downstream of a pollution source in which a pollutant remained detectable. A study of the Ohio River, which is used as a supply of municipal water, showed that 13 major contaminants were present, one of which, bis-(2-chloroisopropyl) ether, was probably derived from an industrial effluent 150 miles away.[83] Other pollutants in the same river were t-butyl mercaptan (4.6 p.p.b.), cresols (1.3 p.p.b.), 2,4-dichlorophenol (6.6 p.p.b.), $C_7$, $C_8$, and $C_{11}$ n-alkanoic acids, and also butyric and benzoic acids; these last two were from a known industrial discharge.[101] By monitoring concentrations at two stations, it was established that the n-acids were biodegradable but that t-butyl mercaptan and 2,4-dichlorophenol were not.

Phthalate esters have been detected in aquatic organisms[102] and in concentrations up to 100 mg $l^{-1}$ in a polluted river water.[82] There is little information on the fate of phthalate esters in the aquatic environment, but they are known to be associated with soil fulvic acids, which could act as a transport and mobilizing agent in water.[103]

The organic composition of domestic waste water, in terms of both general class analysis and specific compounds, has been reviewed.[14] The difficulty of analysing waste-water effluents was emphasized by the observation that general analytical methods which revealed 80% of the organic constituents of the treatment plant influent only revealed 30% of the constituents in the effluent from conventional treatment involving sedimentation and biological oxidation. The effluent consisted of a particulate fraction (12.9 mg $l^{-1}$ carbon), in which protein (2.7 mg $l^{-1}$ carbon) and carbohydrate (1.4 mg $l^{-1}$ carbon) were the major identifiable components, and a soluble fraction (14 mg $l^{-1}$ carbon), in which soluble acids (1.65 mg $l^{-1}$) and anionic surfactants (1.4 mg $l^{-1}$ carbon) were the major identifiable groups. Recent studies have shown that

99 W. J. Weber and J. P. Gould in 'Organic Pesticides in the Environment', Advances in Chemistry Series, No. 60, Amer. Chem. Soc., Washington, 1966.
100 H. M. Gomaa and S. D. Faust in 'Organic Compounds in Aquatic Environments', ed. S. D. Faust and J. V. Hunter, Dekker, New York, 1971, pp. 341—376.
101 F. K. Kawahara, *Environ. Sci. Technol.*, 1971, **5**, 235.
102 F. L. Mayer, D. L. Stalling, and J. L. Johnson, *Nature*, 1972, **238**, 411.
103 G. Ogner and M. Schnitzer, *Science*, 1970, **170**, 317.

cutin, a lipid biopolymer, was a significant component of sewage sludge, constituting 12—28% of the organic matter.[104]

Increasing concern for the effect of phosphorus in eutrophication has led to the use of nitrilotriacetic acid (NTA) to replace phosphates in detergents. The early literature concerning the biodegradability of NTA during sewage treatment and its possible toxicity to fish and other aquatic animals was reviewed in 1971.[105] No information was then available on the effect of NTA on bottom deposits. It has now been shown that NTA, at concentrations of 1 mg $l^{-1}$, will mobilize heavy metals from lake[106] and river[107] sediments, and it was also concluded that NTA would release metal ions from humic complexes and thus change the transport patterns, equilibria, and biological activity of metal ions.

Two related compounds, ethylenediaminetetra-acetic acid (edta) and diethylenetriaminepenta-acetic acid (dtpa), have also been used as complexing agents in detergents and pulp bleach processes, respectively. Using a g.l.c. analysis technique suitable for all three compounds, at concentrations down to 10 $\mu g\ l^{-1}$ for edta and dtpa, it was shown that edta was not degraded during a sewage treatment that removed 80% of NTA.[108] Other analytical methods for determination of NTA and edta have been summarized recently.[109]

The photodecomposition of some industrial pollutants has been studied. Polynuclear aromatic hydrocarbons have a low solubility in water but are concentrated on suspended solids such as calcareous material and silica. The photodecomposition by white light of 3,4-benzpyrene (50) adsorbed on calcium carbonate was shown to follow first-order kinetics with respect to the substrate, with a rate-constant corresponding to a $t_{1/2}$ of 200—250 h.[110] Mono- and di-meric complexes of ferric iron with NTA are decomposed by sunlight under natural water conditions, giving $CO_2$, $CH_2O$, and iminodiacetic acid, with a quantum yield of 0.01—0.02;[111] it was suggested that such decomposition may limit the accumulation of NTA in lakes.

**Organic Mercury Derivatives.**—Mercury enters the aquatic environment either in inorganic form as industrial waste or in organic form as fungicide used for seed dressing or in the paper industry. Investigations of the conversion of inorganic into organic forms of mercury in the aquatic environment have been reviewed.[112] These studies have shown that both methylmercuric ion and dimethylmercury are synthesized from inorganic mercury by bacteria in the surface sediment, the former being produced under aerobic conditions

104 P. E. Kolattukudy and R. E. Purdy, *Environ. Sci. Technol.*, 1973, **7**, 619.
105 N. S. Thom, *Water Res.*, 1971, **5**, 391.
106 C. D. Gregor, *Environ. Sci. Technol.*, 1972, **6**, 278.
107 V. Zitko and W. V. Carson, *Chemosphere*, 1972, **1**, 113.
108 L. Rudling, *Water Res.*, 1972, **6**, 871.
109 K. L. E. Kaiser, *Water Res.*, 1973, **7**, 1465.
110 J. B. Andelman and M. J. Suess in 'Organic Compounds in Aquatic Environments', ed. S. D. Faust and J. V. Hunter, Dekker, New York, 1971, pp. 439—468.
111 T. Trott, R. W. Henwood, and C. H. Langford, *Environ. Sci. Technol.*, 1972, **6**, 367.
112 J. Gavis and J. F. Ferguson, *Water Res.*, 1972, **6**, 989.

and the latter being dominant in anaerobic conditions when sulphide ion was absent. The formation of dimethylmercury by a methanogenic bacterium was shown to proceed by transfer of a methyl group from the $Co^{3+}$ atom of methylcobalamine (51). Under aerobic conditions the reduced form of certain

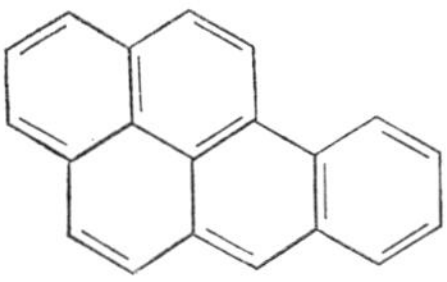

(50)

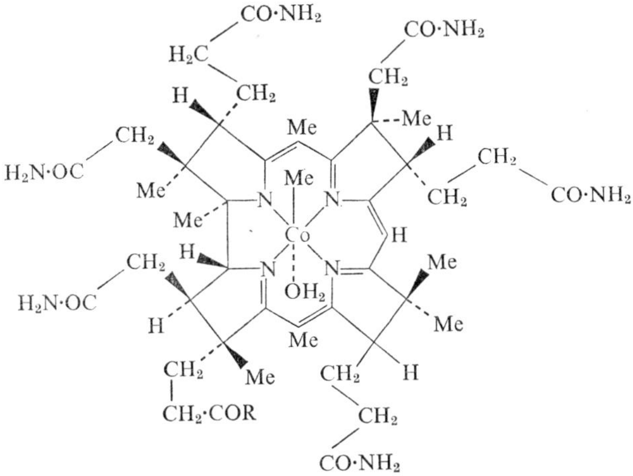


(51)

enzymes catalysed the reaction at low mercury concentrations, but at higher mercury concentrations a non-enzymatic reaction requiring anaerobic conditions occurred. Methylmercuric compounds are lipid-soluble and thus accumulate in the food chain, while dimethylmercury is volatile and can escape into the atmosphere. Variation of pH in the region 5—9, typical of natural waters, did not affect overall methylation but did alter the relative proportion of mono- and di-methylmercury, the latter being favoured at high pH.[113] When $^{203}HgCl_2$ was incubated under aerobic conditions with a lake sediment, the quantity of methylated mercury increased at first and then decreased as methane and $Hg^0$ were formed. When the culture was transferred to a growth medium containing methylmercuric salts, degradation of the latter occurred, and four isolates were obtained from this mixed culture which degraded methylmercury in pure culture.[114] Twenty-one bacterial isolates obtained from environmental samples were capable of demethylating methylmercury under both aerobic and anaerobic conditions.[115]

Phenylmercury compounds have also been introduced into the aquatic environment, and the kinetics of acidolysis and demercuration of both phenyl- and methyl-mercurials have been investigated.[116] Acid cleavage of

113 T. Fagerström and A. Jernelöv, *Water Res.*, 1972, **6**, 1193.
114 W. J. Spangler, J. L. Spigarelli, J. M. Rose, and H. M. Miller, *Science*, 1973, **180**, 192.
115 W. J. Spangler, J. L. Spigarelli, J. M. Rose, R. S. Flippin, and H. M. Miller, *Appl. Microbiol.*, 1973, **25**, 488.
116 N. L. Wolfe, R. G. Zepp, J. A. Gordon, and G. L. Baughman, *Chemosphere*, 1972, **1**, 273; 1973, **2**, 147.

these compounds occurs according to reaction (1).

$$R_2Hg + H^+ \rightarrow RH + RHg^+ \quad (R = Ph \text{ or } Me) \quad (1)$$

The half-life of reaction (1) for diphenylmercury is 8 days at 25 °C and pH 4, and for dimethylmercury $t_{1/2}$ is 33 years at 25 °C and pH 5. Cleavage of diphenyl- and dimethyl-mercury by inorganic mercuric ion species is dependent on the concentration of reactants and pH, and could be significant in acidic waters for diphenylmercury.

While dimethylmercury and methylmercuric compounds do not undergo photodecomposition in sunlight,[116] quantum yields of 0.2—0.3 have been observed for photodecomposition of phenylmercurials with light of >290 nm. The quantum yields were not lowered by quenching processes in the environment, and indicate half lives in sunlight of 8.5 and 17 h for di- and mono-phenylmercury compounds, respectively.[117]

Conversion of phenylmercuric species into methylmercuric species by bacterial action may be an important mode of decomposition of the former. The reaction is reported to occur more rapidly than the formation of methylmercuric compounds from inorganic mercury.[112]

## 7 Stability of Organic Matter in Aquatic Environments

Most of the carbon in the earth's crust has cycled through organisms and plants, thus becoming incorporated into thermodynamically unstable but long-lived structures. Carbon forms the link in the interaction between the inorganic environment and living organisms. Inorganic geochemistry is dominated by equilibrium processes, and most reactions are rapid, so that equilibria are established within a short time span (in geological terms). The equilibrium nature of such systems allows prediction of the stable ionic components from the pH, redox potential, pressure, and temperature. Most organic products of organisms are thermodynamically unstable and those products which escape biodegradation, becoming incorporated in sediments, undergo diagenesis, which leads to gradual equilibration of the sedimentary organic matter. The lack of equilibrium in the latter is manifest in compounds with different oxidation states of carbon in one molecule, whose disproportionation is prevented by slow kinetics, and also in the co-existence of mixtures of compounds with different oxidation state.

The equilibrium composition of multiphase systems of known elemental composition can be calculated from the chemical formula and free-energy of formation of each compound, since the total free-energy of a system is a minimum at equilibrium. In the case of a ternary system such as carbon–hydrogen–oxygen, graphical methods may be used to display the results. Calculations of the equilibrium balance in liquid systems of C, H, O, and N, to determine the quantities of organic compounds in aqueous solution at

[117] R. G. Zepp, N. L. Wolfe, and J. A. Gordon, *Chemosphere*, 1973, **2**, 93.

equilibrium, showed that none existed in significant concentration.[118] In connection with pollution control this negative result implies that any organic compound, under the influence of a suitable catalyst, can be broken down into $CO_2$, $CH_4$, $H_2O$, and $H_2$ or $O_2$. When nitrogen is present, $N_2$ and $HNO_3$ or $NH_3$ are also formed.

**Steady-state Model of the Environment.**—Natural waters are systems open to their environment, and if input is balanced by output in such a system a steady-state condition is obtained and the system remains unchanged with time. Within a body of water energy-rich bonds are produced by photosynthesis, thus distorting the thermodynamic equilibrium. Bacteria and other organisms causing respiration tend to restore equilibrium by catalysing the decomposition of the unstable products of photosynthesis. The steady state has been chemically characterized by the following stoicheiometry[76] (on the basis of N:P ratios in marine plankton):

$$106CO_2 + 16NO_3^- + HPO_4^{2-} + 122H_2O + 18H^+ + \text{(trace elements; energy)}$$

$$R \uparrow \downarrow P$$

$$\underbrace{C_{106}H_{263}O_{110}N_{16}P_1}_{\text{algal protoplasm}} + 138O_2$$

The steady-state balance for an open system is characterized by:

$$I + P \simeq R + E$$

where $I$ and $E$ are the rate of import and export, respectively, of organic matter, $P$ is the rate of photosynthetic production, and $R$ the rate of heterotrophic respiration. A disturbance of the balance between photosynthesis and respiration leads to chemical and biological changes which constitute pollution. When $P > R + E - I$, a progressive accumulation of algae leads to an organic overloading of the receiving waters, while dissolved oxygen may be exhausted if $R > P + I - E$, causing formation of $CH_4$. In a stratified lake, a vertical separation of $P$ and $R$ results from the fact that algae are only photosynthetically active in the euphotic zone; algae that have settled serve as food for the heterotrophic organisms in the deeper layers. In a river a longitudinal separation of autotrophic and heterotrophic organisms is found; algae produced in a nutrient-rich area are transported downstream, where they are decomposed by heterotrophic organisms. The methane produced by decomposition in the absence of oxygen, being only slightly soluble in water, can escape into the atmosphere and thus provide a way for a body of water to rid itself of extraneous organic matter. However, some of the methane carbon is recycled by bacteria that oxidize methane. The action of particulate clay

[118] M. O. Dayoff in 'Organic Compounds in Aquatic Environments', ed. S. D. Faust and J. V. Hunter, Dekker, New York, 1971, pp. 1—28.

minerals in enhancing methane oxidation in the water column and the inhibitory effect of certain living algal cells has been reported.[119]

## 8 Stable Carbon Isotope Distribution

The major steps in carbon isotope fractionation during photosynthesis have been reviewed by Degens,[120] and biological cycling of the carbon isotopes is discussed by Smith (this volume, Ch. 3). Organic matter in Recent sediments has about the same isotopic composition as organisms living in the environment of deposition; in freshwater sediments mean $\delta^{13}C$ values around $-25‰$ are found.[120] Relative to plant organic matter as a whole, lipids are usually depleted in $^{13}C$. A similar result has been observed in sediments from the Dead Sea, in which $\delta^{13}C$ values of $-23.8$ to $-24.3$ were observed for the total organic carbon, compared with $-27$ to $-28‰$ for the hydrocarbon fraction.[43] The pigments from these sediments showed a similar depletion in the heavier isotope, relative to the total organic matter. Since the isotope ratios of the total sedimentary organic matter were very similar to those of the particulate matter in the water and the organic matter in the surrounding soil, it was not possible to use the values as a criterion of origin of sedimentary organic matter. In the sediment profile from Lake Biwa the isotopic composition of the organic carbon became heavier as the organic carbon content of the sediment increased.[121] The isotopic variation was attributed to the kinetic isotope effect and the temperature at which the plankton grew. Variations in isotope composition down the profile were thus interpreted in terms of climatic change, a hypothesis which was supported by palynological studies.

119 T. L. Weaver and P. R. Dugan, *Water Res.*, 1972, **6**, 817.
120 E. T. Degens in 'Organic Geochemistry. Methods and Results', ed. G. Eglinton and Sister M. T. J. Murphy, Springer-Verlag, Heidelberg, 1969, pp. 304—329.
121 N. Nakai, *Japan. J. Limnol.*, 1973, **34**, 89.

# 3
# Environmental Organic Chemistry of Bogs, Marshes, and Swamps

BY P. H. GIVEN

## 1 Introduction: Characteristics of Wetland Environments

From the definition given in the Shorter Oxford Dictionary, it would appear that the words 'marsh', 'swamp', 'bog', 'fen', 'mire', and 'slough' are essentially synonymous in everyday speech. 'Mire' and 'slough' are no longer commonly used in lay speech, and 'fen' has special associations in England, though not elsewhere. The common factor in the ideas conveyed by all six words is that they refer to waterlogged habitats for higher plants and that they are soft and muddy to walk in.

Wetlands have a number of characteristics in common:

(i) The plant communities tend to be simple, because relatively few species of plants can tolerate a waterlogged habitat.

(ii) The rates of diffusion of oxygen and nitrogen into the soil are restricted, compared with the rates when inter-particle spaces are filled with gases rather than water.

(iii) Rates of microbial decay of plant debris are reduced, so that usually (though not universally) the soil is organic-rich or a peat, and contains fragments of tissue easily recognizable under the microscope.

(iv) Transport of soluble materials is possible by diffusion in water or by bulk flow; such mechanisms are not possible in normal mineral soils.

Although they have these features in common, wetlands include a diverse range of distinct ecosystems that differ in the nature of the plant and animal communities inhabiting them, in their hydrology, and in the typical levels of pH, ionic strength, and productivity of biomass. Bellamy[1] uses the word 'mire' as a generic term for any kind of peat-forming ecosystem, and then classifies mires into a number of hydrologic types. Other classifications are ecological (*e.g.* that of Smith[2]). There are, of course, many transitional or

[1] D. J. Bellamy, Proceedings of the Third International Peat Congress, National Research Council of Canada, Quebec, 1968, pp. 74—79.

intermediate types of wetland, but the basic distinctions are well expressed by Smith[2]:

> 'The bog, marsh, and swamp represent late fill-in stages in lake succession, but swamps also develop on the flood plains of the larger river systems of southern North America. Each is distinctively different. Bogs, confined to northern regions, are characterized by blocked drainage conditions, an accumulation of peat, cushion-like vegetation, a marginal semi-floating mat of plant growth, and acidic conditions, created largely by *Sphagnum* moss. Only those plants tolerant of acidic conditions occupy the bog. Life is restricted in the number of plant and animal species, but the organisms present are often abundant. Marshes, on the other hand, are wetlands in which the grass life-form is dominant, while swamps are wooded. Both may range from deep to shallow water, and both embrace a richness and diversity of life that is hard to equal in other temperate communities.'

There is, of course, an implication that if the drainage of bogs is blocked, so also is the inflow of nutrients and other ions in solution, and materials dissolved in rainwater are important to the life of the plants. A common site for a bog is a glacial 'kettle', a depression ground out of hard rock by the flow of glaciers during the Ice Ages. *Sphagnum* spp., the predominant plant type in bogs, are non-vascular, that is, their members do not have woody parts with lignified cell walls. At the present time, the overwhelming majority of peat-forming areas of the world are in northern temperate latitudes (mostly in the U.S.S.R. but also in Scandinavia, Poland, Ireland, Canada, and the northern U.S.A.).[3] A considerable proportion of these peatlands are bogs as characterized by Smith in the passage quoted.

Marshes and swamps do have an input of water-borne substances from moving ground waters, and their margins are not as well defined as those of bogs often are. Salt marshes, in coastal areas, have been of great interest to many botanists and ecologists, at least partly because their annual productivity of biomass is second only to that of tropical rain forests (it is greater than that of the heavily fertilized deep loam soils of the wheatlands of the American mid-west). In sub-tropical regions, areas that in higher latitudes would be salt marshes are instead mangrove swamps. The pH of fresh and saline swamps and marshes (typically 4.5—7) is higher than that of bogs (often 2—4).[4]

Peatlands have been variously estimated to cover about 0.4%[3] and 0.7%[5] of the land surface of the earth. Peat has been extensively used as a fuel in

[2] R. L. Smith, 'Ecology and Field Biology', Harper and Row, New York, 1966, pp. 182—191 and 215—220.

[3] A. M. Matveev, M. I. Neistadt, and A. S. Olenin, Proceedings of the Third International Peat Congress, National Research Council of Canada, Ottawa, 1968, p. 382.

[4] S. M. Manskaya and T. V. Drozdova, 'Geochemistry of Organic Substances', transl. L. Shapiro and I. Breger, Pergamon, Oxford, 1968.

[5] M. N. Nikonov, *Byoll. MOIP see Geol.* 1948, **23**, 93 (puoted in ref. 4, p. 29).

regions that lack coal and oil. Bogs, marshes, and swamps are often regarded as wastelands that need to be 'reclaimed' for agricultural use by draining, liming, and fertilizing. They are in fact ecosystems of great interest that should be managed and conserved. However, it must be admitted that inhabitants of countries where a large fraction of their land area is covered by wetlands (Finland 19%, Ireland 17%, and Britain 10%)[3] may take a different view.

It will be clear from the descriptions above that bogs, marshes, and swamps will have distinctive chemistries to match their differing ecologies. Moreover, in any type of waterlogged habitat, a variety of circumstances may determine the extent to which the input of plant debris is microbiologically degraded. The extent of decomposition has an influence on the colloidal characteristics of the peat or soil material, the content of oxygen functional groups that are capable of ion-exchange processes and of complexing metals by chelation, and the permeability of the sediment to water flow and to diffusion in solution. It has often been the case that particular types of investigation have been made in only a few wetland habitats. Generalizations are therefore often hazardous.

## 2 Some Ecological Aspects

A survey of a number of freshwater peat mires in Europe indicated a rate of accumulation of 0.1—1.6 mm $(year)^{-1}$, the mean being 0.4 mm $(year)^{-1}$.[6] On the basis of average depths of peat and $^{14}C$ dating of the basal sections,[7,8] the average rate of accumulation in the mangrove swamps of southern Florida has been around 1 mm $(year)^{-1}$. Rates of 1—10 mm $(year)^{-1}$ have been noted in Connecticut salt marshes.[9,10] These rates are higher by several orders of magnitude than those observed for sediments on the continental shelves, where in any case the content of organic matter in the sediment is lower than in mires.

The composition of a freshwater mire might commonly be 85—90% water, 9—14% organic matter, and 1—2% inorganic materials.[11] In marshes and in saline swamps there may be considerably higher proportions of inorganic materials (up to 60—70% on a dry basis),[12,13] depending on the geological

6 G. Grosse-Brauckmann and D. Puffe, Transactions of the 8th International Congress on Soil Science, Bucharest, 1964 (Pub. 1967), vol. 5, p. 635.
7 W. G. Smith, Ph.D. thesis, 'Sedimentary Environments and Environmental Changes in the Peat-forming Area of South Florida, The Pennsylvania State University, University Park, 1968.
8 W. Spackman and P. H. Given, unpublished data, 1971.
9 A. L. Bloom, 'Coastal Morphology of Connecticut', Final Report, Office of Naval Research, Contract No. 401(45) Task No. 388-065, 1967.
10 T. G. Siccama and E. Porter, *Bioscience*, 1972, **22**, 232.
11 P. H. Given and C. H.Dickinson, in 'Soil Biochemistry', ed. E. Paul and A. D. McLaren, Marcel Dekker, New York, 1974, vol. 3, p. 123.
12 V. J. Chapman, 'Salt Marshes and Salt Deserts of the World', Leonard Hill (Books) Ltd., London, 1970, pp. 80—108.
13 D. J. Casagrande and P. H. Given, *Geochim. Cosmochim. Acta*, 1974, **38**, 419.

setting, but this is still an organic-rich growth medium for plants compared with normal mineral soils. (Unfortunately, many excellent studies of the ecology of marshes note that the plants are growing in peat or organic-rich soil but give no quantitative data for the organic content.)

Thus a very significant characteristic of wetlands is that there is a marked inefficiency in the recycling of carbon. Few data are available to show the magnitude of this inefficiency. Considering only the productivity of the aerial parts of plants (*i.e.* excluding productivity of root material), estimates of about 10% of total annual production preserved as peat have been made.[6,11] The inefficiency of recycling is discussed further below.

It will be noted that coastal marshes and swamps show the highest annual rate of increase in depth and that they contain also the highest proportion of inorganic constituents. The two facts are no doubt related. The tide, covering the marsh twice a day, transports and dumps particulate matter, which accumulates with the organic debris. There is usually also some input of particulate matter and soluble nutrients from the landward side, particularly in cultivated areas.

The most common plants growing in salt marshes in many parts of the world are the rushes *Spartina alternifolia* and *S. patens*, and the cat-tail, *Typha* spp. The contents of dissolved phosphate and nitrogen forms in the waters may be modest or quite high.[14,15] Phosphate is adsorbed on clay minerals or inserted between the layer planes of the clays, but it can be released when required by the plants; *Spartina* has been described as a nutrient pump, raising phosphate from the sediment to the surface of the marsh.[15] Detritus from the plants is colonized by micro-organisms, which may in addition use nitrogen from the waters in their metabolism. In this way partly decayed leaf material may be richer in protein than fresh plant tissue.[14,16]

Thus the detritus of *Spartina* is an important element in food chains in salt-marsh ecosystems. Moreover, a comparatively small part of the detritus is consumed within the salt marsh, which, therefore, contributes substantial amounts of mixed nutrients to a neighbouring estuary.[14,16] In creeks draining the Sapelo marshes in Georgia (U.S.A.) the content of organic detritus in the water ranges from 2 mg $l^{-1}$ at mid-flood tide to 20 mg $l^{-1}$ at mid-ebb. On an ash-free basis this detritus is 24% protein as compared with 10% in living leaves of *S. alternifolia*. Hence the bacteria-rich detritus is a better food source for animals than is *Spartina*, the primary producer in the ecosystem.[16]

The detritus and peat are ecologically important in other respects. Peat has very considerable water-holding capacity (see also below) so that, even in drought conditions, enough water is retained for healthy growth of the plants. Decomposition of the surface organic matter produces much $CO_2$, which reduces the somewhat alkaline pH of sea water and thereby somewhat

[14] C. W. Keefe, *Contrib. Marine Science*, 1972, **16**, 163.
[15] R. J. Reimold and F. C. Daiber, *Hydrobiologia*, 1970, **36**, 361.
[16] E. P. Odum and A. A. de la Cruz, *Amer. Inst. Biol. Sci. Bull.*, 1963, **13**, 39; *Publ. Amer. Assoc. Adv. Sci.*, 1967, **83**, 383.

increases the solubility of Fe and Mn; *Spartina* has an unusually high requirement for iron.[17] Colloidal organic materials in the peat can adsorb and make available a variety of ions necessary for plant growth. The various physicochemical processes in peats evidently vary considerably with the seasons; wide seasonal fluctuations of pH, oxygen content, and salinity of the waters in a salt marsh have been noted,[18] as well, of course, as variations in temperature and productivity of vascular plants and algae.[19]

No information has been found on the nitrogen and phosphorus economy of the peats in coastal swamps, but it seems a fair inference that there are inputs similar to those into marshes. There is evidence that leaves shed from mangrove trees in the coastal swamps of southern Florida play essentially the same role in providing nutrition for animals in local bays and estuaries as do *Spartina* leaves in a salt-marsh environment.[20,21]

The chemical ecology of freshwater mires has been much less studied. The input of nutrients is more variable than in saline marshes and swamps, and will depend on, *inter alia*, local topography, and the nature of the rocks being eroded in any neighbouring hills or mountains.

The nitrogen content of freshwater mires, on the dry basis, is commonly 2—4%,[13,22,23] but much the greatest part of this is organic; for example, 97—99% of the total N in various marshes in Poland is organic, and 89—96% in bogs.[24] Of the organic nitrogen, 30—60% is removed from complexes by strong acid hydrolysis as amino-acids, the balance being non-hydrolysable.[13,23] Nitrate contents around 1—10 p.p.m. have been reported in some Russian marshes,[22] and 0.008—0.1 p.p.m. N as $NO_3^-$ in the surface waters of freshwater marshes in the Everglades of Florida.[25] The latter study also reports concentrations of 0.03—0.1 p.p.m. N as $NH_3$. How much of the nitrogen in peats is available to the microflora and macroflora is a matter of controversy.[26]

17 J. M. Teal and M. Teal, 'Life and Death of a Salt Marsh', Little Brown and Co., Boston 1969.
18 B. Phleger and J. S. Bradshaw, *Science*, 1966, **154**, 1551.
19 J. W. Day, W. G. Smith, P. R. Wagner, and W. C. Stowe, 'Community Structure and Carbon Budget of a Salt Marsh and Shallow Bay Estuarine System in Louisiana', Publication No. LSU-SG-72-04, Center for Wetland Resources, Louisiana State University, 1973.
20 W. E. Odum, 'Pathways of Energy Flow in a South Florida Estuary', Sea Grant Tech. Bull. No. 7, University of Miami, 1971.
21 D. C. Tabb, D. L. Dubrow, and R. B. Manning, 'The Ecology of Northern Florida Bay and Adjacent Estuaries', Tech. Series Florida Board of Conservation, 1962, vol. 34, pp. 1—79.
22 S. T. Voznuk, T. Y. Korobchenko, and N. N. Skochinskaya, *Soviet Soil Sci.*, 1964, No. 1, p. 12; A. M. Krasoinskaya and V. G. Timofeev, *Mater. Probl. Geol. Pozdnego Kaiozoya*, 1969, p. 161.
23 V. E. Rakovskii and I. A. Pal'min, *Torf. Prom.*, 1965, **42**, 23 (*Chem. Abs.*, 1965, **63**, 17 082).
24 F. Maciak, A. Maksimow, and S. Liwski, Proceedings of the Second International Peat Congress (Leningrad 1963), ed. R. A. Robertson, Her Majesty's Stationery Office, Edinburgh, 1968, vol. II, p. 919.
25 P. J. Gleason, 'Chemical Quality of Water in Conservation Area 2A and Associated Canals', Tech. Publ. No. 74-1, Florida Flood Control, 1974.
26 H. van Dijk, J. van der Boon, and P. Bockel, Proceedings of the Third International Peat Congress, National Research Council of Canada, Ottawa, 1968, p. 334.

The few reported studies of aerobic and anaerobic nitrogen-fixing bacteria in peats (see Given and Dickinson,[11] and Mishustin and Shil'nikova,[27] and references therein) indicate that they are scarce or absent, particularly in bogs. *Azotobacter* does not tolerate a pH below about 5.6. Since freshwater marshes can have quite high productivities, an effective means of obtaining nitrogen must be inferred. Blue-green algae are the dominant nitrogen fixers in rice paddies,[27] but their possible role in mires has not been studied.

Identification of micro-organisms in any natural habitat is always an uncertain means of identifying important microbial processes in the habitat, because even if organisms are present, their rate of activity may be low. In the case of nitrogen fixation, exposure of soils to $^{15}N_2$ or to acetylene provides an excellent way of determining the actual rate at which the process occurs under natural conditions.[28] The nitrogenase enzyme catalyses the reduction of acetylene to ethylene, the production of which can be measured easily by gas chromatography; this technique is simpler, more sensitive and less expensive than the use of $^{15}N$.

Nitrogen fixation in a marsh/bog complex has recently been estimated by use of the acetylene procedure.[29] One study site (in North-west Scotland) was composite in that the outer margin received soluble materials from a stream and was populated by typical marsh plants; the centre was essentially stagnant and populated by *Sphagnum* spp., *Erica tetralix*, and *Calluna vulgaris* (and other species), and there were also transitional zones. A similar site in Perthshire was also studied. Samples of peat were taken from 15—20 cm below the surface.

In the outer marsh, the production of ethylene was $1.8 \times 10^{-9}$ mol (ml peat)$^{-1}$ day$^{-1}$, and the figure steadily decreased to $0.10 \times 10^{-9}$ at the centre. These figures correspond approximately to the range 1.2—0.07 g $N_2$ fixed m$^{-2}$ year$^{-1}$. Addition of glucose greatly increased the rate. Slow anaerobic fixation was observed with peat from 25 cm depth. Exposure to light had no effect on fixation rates, so presumably algae were not the organisms responsible. The rates observed were low, but could significantly supplement inputs from precipitation. Further studies of this type in valley marshes and swamps would be very illuminating.

There are indications that although the content of available nitrogen in some marshes and swamps may be reasonably high, the phosphate level is low.[30,31]

A detailed study of the history of a bog in Illinois well illustrates how bogs in general came to have their characteristic hydrology and chemistry.[32] The

27 E. N. Mishustin and V. K. Shil'nikova, Biological Fixation of Atmospheric Nitrogen', The Pennsylvania University Press, University Park, Pa. and London, 1971.
28 W. A. Rice and E. Paul, *Canad. J. Microbiol.*, 1971, **17**, 1049.
29 G. J. Waughman and D. J. Bellamy, *Oikos*, 1972, **23**, 353.
30 O. S. Shchukin, *Nauk. Pr., L'viv. Sil's kogospod. Inst.*, 1970, **27**, 42 (*Chem. Abs.*, 1972, **77**, 86 914).
31 A. F. Borovkova and V. I. Troitskii, *Uch. Zap. Tomsk. Gos. Pedagog. Inst.*, 1969, No. 27, p. 42 (*Chem. Abs.*, 1972, **77**, 74 082).
32 M. R. McComas, J. P. Kempton, and K. C. Hinkley, *Environ. Geol. Notes*, 1972, No. 57, 27 pp.

statement, 'peat is a nutritionally impoverished medium for microbial and plant growth' was made without qualification,[33] but in fact refers only to peats accumulating in raised bogs (and was made by authors interested in converting bogs to agricultural use). The bogs studied were in the west of Ireland. No evidence of nitrification (oxidation of $NH_3$ to $NO_3^-$) could be detected unless the bog was drained, limed (to raise pH), and left for several years. It appears that nitrate is the only form of nitrogen usable by many agricultural crops. Other authors[34] have found the content of Na, K, and $NO_3^-$ in bogs to increase appreciably during heavy rainfall; a significant source of phosphate for these environments was wind-borne pollen.

## 3 Water in Peats

We are concerned in this chapter with waterlogged terrestial habitats, in which the medium for growth of higher plants is of high organic-matter content. In mineral soils clay minerals play a vital role in concentrating, by adsorption and ion exchange, the various ions and trace elements needed by the soil biota. In organic-rich soils this function is largely taken over by organic matter. Transport of ions and gaseous $O_2$, $N_2$, and $CO_2$ must be by diffusion in the water medium or by bulk flow of the water. Therefore, the nature of the association of water with organic matter and the possibilities of bulk flow are significant parts of the environmental chemistry of wetlands.

Typically, bogs have an impermeable bottom layer of hard rock, so that their water circulation is isolated from ordinary ground-water movement. Where the degree of decomposition of the sediment is low, the peat consists of a tangled mass of stems and leaves of *Sphagnum*, and water movement is comparatively easy. If the degree of decomposition is high, the deposit consists largely of a hydrophilic colloid, a gel, which is almost impermeable to bulk flow. Provided that precipitation of rain and snow is adequate in amount and in nutrient content, it is a property of bogs that their upper surface rises above the surrounding terrain, the bog water-table rising at the same time. It has been shown that the maximum height of a raised bog is proportional to the square-root of the annual rainfall and to the square-root of the diameter of the bog, and the reasonableness of this curious behaviour has been demonstrated theoretically.[35]

That the circulation of water in raised bogs is poor has been shown by some interesting observations of tritium distributions in bogs in northern Minnesota.[36] The *Sphagnum* in the peats was not extensively decomposed. In recent years atmospheric moisture has been enriched in $^3H$ as a result of the testing of nuclear weapons. In 1965, the surface peat showed 1386 tritium

[33] E. Küster and J. J. Gardiner, Proceedings of the Third International Peat Congress, National Research Council of Canada, Ottawa, 1968, p. 314.
[34] S. Carausu and T. Boisteanu, *Anal. Sti., Univ. "Al. I. Cuza" Iasi, Sect. 2a*, 1970, **16**, No. 1, p. 163 (*Chem. Abs.*, 1971, **74**, 102871).
[35] F. E. Wickman, *Geol. Fören Förhandl.*, 1951, **73**, 413.
[36] E. Gorham and R. H. Hofstetter, *Ecology*, 1971, **52**, 898.

units [1 T.U. = $3.24 \times 10^{-12}$ Ci (liter water)$^{-1}$] and 11 T.U. at a depth of about 3.5 m. A more detailed profile determined in 1969 showed low activity at the surface (320 T.U.) and a sharp drop to about 30 T.U. at 1.8 m. Values remained at about this level to 3.5 m, where there was a further sharp drop to 8 T.U., which was a typical level before the testing of nuclear bombs was started.

It appears that $^3$H-enriched precipitation in the previous 10 years or so had mostly run off the surface of the bog or run out laterally through the upper layers of peat or had been lost by evapotranspiration of the moss plants. However, a little did penetrate by diffusion down to 3.5 m; at this depth there was an interface with a more highly decomposed and less permeable underlying sedge peat. Surely we may infer that if in 5—10 years there is not much mixing of water in the top 1.5 m or so with that lower down, then diffusion of oxygen, nitrogen, and ionic species in solution in bogs must also be slow; this bears on possible microbial activity in lower levels of peat.

In marshes and swamps, where there are inputs from ground waters, the situation will be more complex and has been little studied. In tidal marshes and swamps there is the possibility of partial drainage at low tide and replenishment with nutrient- and oxygen-laden waters at high tide. The southern Florida peninsula is underlain by a number of quite porous limestone formations, which contain the aquifers from which the city of Miami obtains its water supply. In the wet season, therefore, it is likely that bulk flow of surface water downwards through the peat into the underlying limestone occurs. The observation of sharply increasing salinity with depth in the peat in brackish areas in the rainy season is consistent with this suggestion.[37] In other areas the rock underlying the peat may be less permeable.

A number of studies have shown that water is held in peat in several distinct ways (for a fuller review of this topic, see refs. 11 and 38). Any plant tissue in peats is highly swollen, and water will tend to be immobilized in cell walls and within the lumina of intact cells. In addition, the fine-grained amorphous material will have colloidal properties. Integration of desorption isotherms indicates a total evaporation energy of more than 15 kcal mol$^{-1}$ (*cf.* 9.7 kcal mol$^{-1}$ for pure water).[38] Studies of the diffusion of a variety of radioactive tracers in peats[39–41] permit the distinguishing of free intercellular water from structural gel-bound or immobilized water; the latter has no dissolving power for ions. These conclusions have been confirmed and extended

[37] P. H. Given and C. Exarchos, unpublished observations, 1971.

[38] M. P. Volarovich, N. I. Gamayunov, and I. I. Lishtvan, Proceedings of the Fourth International Peat Congress, Otaniemi, Finland, 1972, vol. 4, p. 219.

[39] M. P. Volarovich and N. V. Churaev, Proceedings of the Second International Peat Congress, (Leningrad 1963), ed. R. A. Robertson, Her Majesty's Stationery Office, Edinburgh, 1968, vol. II, p. 819.

[40] M. P. Volarovich, I. I. Lishtvan, and N. V. Churaev, *Kolloid Zhur.*, 1973, **25**, 22 (*Chem. Abs.*, 1963, **59**, 1124).

[41] N. T. Korol, *Prir. Bolot. Metody Ikh Issled., Mater., Vses Soveshch., Sovrem. Puti Metody Issled. Bolot.*, 1964 (Publ. 1967), p. 165 (*Chem. Abs.* 1968, **69**, 108 502).

by studies of electrokinetic potentials, of spin–lattice relaxation and spin-relaxation times determined by n.m.r. methods,[38] and of dielectric behaviour.[42] The dielectric studies resulted in a qualitative evaluation of the energy of interaction of the bound water with its substrate and showed that this water is only weakly structured.

Korol[41] gives for a variety of peat types the 'chemically and physically bound' water as in the range 38—56% and 'strongly bound' water as 9—13%. An increase in the concentration of multivalent ions converts the peat colloids into a coagulated gel structure that immobilizes much water, whereas an increase in [$Na^+$] or [$K^+$] tends to give sols or true solutions.[39] As ions are eluted from a column of peat with deionized water, the flow rate decreases up to ten-fold,[42] with changes in the inter- and intra-aggregate hydrogen bonds.[43]

It is clear that the migration of soluble species in peats will be considerably slower than might be expected from their high water contents. This has relevance to nutritional questions as well as to penetration by substances released by man's activities. The transport of ions in peats has been described as a gel diffusion process.[42] The physical chemistry of water in peat will have extra complexities when high ionic strengths are present.

## 4 Organic–Inorganic Interactions in Peats

**Ion-exchange Behaviour.**—The subject has been reviewed recently.[11] The total ion-exchange capacity of peats is in the range[39,44] 20—200 meq (100 g dry peat)$^{-1}$ (although the measurements are expressed in this way, they are made on undried peat since much of the capacity is irreversibly lost on drying[39]). The ion-exchange capacity varies widely with the ion considered, being highest for $H^+$; certain ions, such as $Cu^{2+}$, may show a greater uptake than $H^+$, but partly covalently bonded complexes are then involved.[45] The greater part of the capacity is due to the carboxy-groups of the humic and fulvic acids, though at high pH exchange with phenolic groups is possible.

Ion-exchange equilibria and kinetics have both been studied.[42,44,46,47] As expected from a consideration of charge/ionic radius ratios, capacity for bivalent ions is greater than for univalent, and exchange capacity increases

[42] P. I. Bel'kevich, L. R. Chistova, and L. F. Strogonova, *Khim. Tverd. Topl.*, 1973, 3, 47 (*Chem. Abs.*, 1971, **75**, 80 660); M. P. Volarovich, N. I. Gamayunov, and K. S. Pantelei, *Kolloid Zhur.*, 1972, **34**, 333.

[43] A. M. Lych and P. N. Davidovskii, *Teplo-Massoperenos, Doklady Vses. Soveshch. 4th*, 1972, **7**, 432 (*Chem. Abs.*, 1973, **78**, 100 282); N. I. Gamayunov, A. M. Lych, and P. Davidovskii, *Inzh. Fiz. Zhur.*, 1972, **22**, 795 (*Chem. Abs.*, 1972, **77**, 79 779); *Khim. Tverd. Topl.*, 1972, **6**, 57 (*Chem. Abs.*, 1973, **78**, 113 745).

[44] P. I. Bel'kevich and L. R. Chistova, Proceedings of the Second International Peat Congress (Leningrad 1963), ed. R. A. Robertson, Her Majesty's Stationery Office, Edinburgh, 1968, vol. II, p. 909.

[45] P. I. Bel'kevich, L. R. Chistova, E. A. Yurkevich, and T. V. Zaitseva, *Vestsi Akad. Navuk Belarus. S.S.R., Ser. Khim. Navuk*, 1972, **4**, 50 (*Chem. Abs.*, 1972, **77**, 112 932).

[46] P. I. Bel'kevich, L. R. Chistova, and L. F. Strogonova, *Vesti Akad. Navuk Belarus. S.S.R., Ser. khim. Navuk*, 1966, **4**, 29 (*Chem. Abs.*, 1967, **66**, 108 616); R. C. Salmon, *J. Soil Sci.*, 1964, **15**, 273.

[47] P. I. Bel'kevich, L. R. Chistova, and L. F. Strogonova, *Vesti Akad. Navuk Belarus S.S.R., Ser. khim. Navuk*, 1971 **4**, 27 (*Chem. Abs.*, 1972, **76**, 36 124).

in the order $Li^+ < Na^+ < K^+$, *etc.*, and $Mg^{2+} < Ca^{2+}$, *etc.* Equilibrium is independent of temperature in the range 25—75 °C, so that the enthalpy change is small or zero, but the replacement of H by cations increases the entropy of the system.[44] Studies[45,48] of desorption and by the use of electrochemical and spectrochemical methods show that a number of ions form chelate complexes, notably $Fe^{2+}$, $Cu^{2+}$, $Zn^{2+}$, and $Mn^{2+}$.

As indicated, the exchange capacity of peats is largely determined by the humic acid contents, which in turn are related to the degree of decomposition of the plant material.[49,50] The contents are generally held to be in the range 5—25% for *Sphagnum* bogs and 30—80% for freshwater marshes inhabited by monocotyledons and for forested swamps.[4,50] However, peats in the reed and sedge marshes and the mangrove swamps of the Florida Everglades are apparently anomalous in that humic acid contents are in the range 4—20% only.[13,51]

A variety of physical methods has been brought to bear recently on the interactions of humic, fulvic, and hymatomelanic acids with cations.[52] The acids were obtained from a Hungarian marsh peat and carefully purified by dialysis. The exchange capacity for $H^+$ was 800—900 meq $(100\ g)^{-1}$ for fulvic and hymatomelanic acids, and 400—600 meq $(100\ g)^{-1}$ for humic acids. Capacities for a series of multivalent cations were appreciably less than these values and ranged down to 50—150 meq $(100\ g)^{-1}$ for $Mg^{2+}$ and zero for $Cr^{3+}$. I.r. and Mössbauer spectroscopy, and e.p.r. measurements, permitted estimates of the partial ionic character of the cation–carboxylate bond; thus bonds formed by $Fe^{2+}$ were 26% ionic in a humate and 37% ionic in a fulvate. Particle weights measured with an analytical ultracentrifuge depended not only on pH but also increased on adding various cations in concentrations 1—10 mmol $l^{-1}$. The molecular weights of the rigorously purified humic acids appear to be no more than a few thousand.

Large amounts of data are available on the concentrations of the common cations in mires.[38,41,53] Except in saline environments, $Ca^{2+}$ and $Mg^{2+}$ are

48 F. D. Ovcharenko and S. A. Gordienko, *Agrokem. Talajtan*, 1969, **18**, 25 (*Chem. Abs.*, 1960, **71**, 52 042); R. S. Solodovnikova, L. A. Skripal'schikova, and V. V. Serebrennikov, *Trudy Tomsk. Gosud. Univ.*, 1968, **192**, 113 (*Chem. Abs.*, 1970, **73**, 59 157); R. I. Davies, M. V. Cheshire, and I. J. Graham-Bryce, *J. Soil Sci.*, 1969, **20**, 65.

49 V. E. Rakovskii, V. Baturo, and L. Pigulevskaya, Proceedings of the Second International Peat Congress (Leningrad 1963), ed. R. A. Robertson, Her Majesty's Stationery Office, Edinburgh, 1968, vol. II, p. 965.

50 I. M. Kurbatov, Proceedings of the Second International Peat Congress (Leningrad 1963), ed. R. A. Robertson, Her Majesty's Stationery Office, Edinburgh, 1968, vol. I, p. 133.

51 A. J. Lucas, 'Geochemistry of Carbohydrates in some Organic Sediments of the Florida Everglades,' Ph.D. Thesis, The Pennsylvania State University, University Park, 1970.

52 B. Lakatos, J. Meisel, G. Mády, P. Vinkler, and S. Sipos, Proceedings of the Fourth International Peat Congress, Otaniemi, Finland, 1972, vol. 4, 341; S. Sipos, E. Sipos, I. Dékány, F. Szántó, and B. Lakatos, *ibid.*, p. 255; S. Sipos, I. Dékány, F. Szántó, and E. Sipos, *Acta Phys. Chem.*, 1972, **18**, 253 (*Chem. Abs.*, 1973, **78**, 109 844).

53 M. P. Volarovich, N. T. Korol, I. L. Lishtvan, A. M. Mamtsis, and N. V. Churaev, *Prir. Bolot. Metody Ikh Issled., Mater., Vses. Soveshch. Sovrem. Puti Metody Issled., Bolot.*, 1964 (Publ. 1967), p. 149 (*Chem. Abs.*, 1968, **69**, 108 503); T. Mornsjo, *Botan. Notiser.*, 1968, **121**, 343 (*Chem. Abs.*, 1969, **70**, 70 265).

the most abundant cations. As would be expected from the descriptions given earlier, bogs, which are of low pH and dependent mostly on rainfall for nutrition, have considerably lower cation concentrations than marshes and swamps. Volarovich,[53] in a study of 288 peat samples, found the correlation coefficient for the dependence of $Ca^{2+}$ concentration on pH to be 0.84, and 0.91 for $Ca^{2+} + Mg^{2+}$. In addition to these basic differences, there are effects of geological environment[54] and of the nature of the plant cover.[55]

**Trace Elements.**—A study of adsorption isotherms shows clearly that humic acids have great affinity for a variety of cations.[56] Nevertheless, reported concentrations of many elements in undisturbed peats are in most cases well below their average abundances in the earth's crust.[11] In any individual marsh or swamp, the concentration of trace elements depends strongly on the nature of the rocks around the basin and underlying the organic sediment, and also on local hydrology; moreover, there are considerable lateral variations in any stratum.[57,58] The situation seems to be that the input to any mire is very variable; it tends to be trapped around the margins of the mire or along stream banks, when the source is external. If the input of elements is moderately large, some may penetrate to the centre of the mire, but may then show a concentration gradient, decreasing with increasing depth. When the underlying bedrock is the source, the vertical gradient will be reversed.

The ecological consequences of this situation have, as far as the author is aware, not been studied. All organisms need small quantities of a variety of elements. One factor limiting the productivity of bogs may be a deficiency of some micronutrients. Saline marshes and swamps presumably can obtain a rich supply of trace elements from sea water. The productivity of some freshwater marshes and swamps and the character of the communities inhabiting them may be determined at least in some portion of the areal extent by availability of micronutrients. Microflora and macroflora do not, of course, require identical elements in identical amounts, and so one type of growth might be limited while the other is not.

It is clear that most mires will have a considerable capacity for trapping cations that may be released by man's activities. Even if these are toxic to

54 V. G. Klingor, I. F. Largin, and M. V. Popov, *Prir. Bolot. Metody Ikh Issled., Mater., Vses. Soveshch., Sovrem. Puti Metody Issled. Bolot.*, 1964 (Publ. 1967), p. 254 (*Chem. Abs.*, 1969, **70**, 49 260).

55 S. N. Tyuremnov and I. F. Largin, *Trudy Gosud. Gidrol. Inst.*, 1966, No. 135, p. 223 (*Chem. Abs.*, 1968, **68**, 53 143).

56 A. Szalay and M. Szilágyi, in 'Advances in Organic Geochemistry 1968', ed. P. A. Schenk and I. Havenaaar, Pergamon, Oxford, 1969, p. 567; N. A. Kanunnikova and V. I. Roslikova, *Uch. Zapl., Dal'nevost. Gosud. Univ.*, 1969, **27**, 108 (*Chem. Abs.*, 1972, **77**, 18 553).

57 F. Ya. Saprykin and A. N. Sventikhovskaya., *Mater. Soveshch. Rabot. Labor. Geol. Organiz.*, 19th Leningrad, 1965, No. 7, p. 95 (*Chem. Abs.*, 1967, **66**, 67 841); S. E. Priemskaya, *Trudy Kalinin. Politekh. Inst.*, 1970, No. 5, p. 8 (*Chem. Abs.* 1972, **76**, 116 064).

58 S. E. Priemskaya, *Pochvovedenie*, 1969, **4**, 59 (*Chem. Abs.*, 1969, **71**, 23 681); I. F. Largin, S. E. Priemskaya, A. N. Sventikhovskaya, and S. N. Tyuremov, Proceedings of the Fourth International Peat Congress, Otaniemi, Finland, 1972, vol. 4, p. 77.

animal life, animals may in fact not be poisoned if the element is tightly bound to insoluble organic matter and not released by the activities of microflora and macroflora. Studies of the trace elements naturally present in marshes show that a widely ranging proportion (10—80%) of certain elements may be mobile and available to organisms, depending on the element and the season of the year.[58]

**Sulphur.**—One of the striking differences between freshwater and saline mires is the sulphur content. For example, the total sulphur content of the peats of the coastal swamps of south Florida is 2—6% (dry basis), while that of freshwater peats of the Everglades is uaually 0.1—0.6%.[59,60] The Carboniferous coals of the Interior Provinces of North America (Illinois, Missouri, Kansas, *etc.*) were, from palaeontological evidence, formed in saline peat swamps and have sulphur contents of 3—6%. The Cretaceous and Tertiary coals of the Rocky Mountain and North Great Plains Provinces were formed in freshwater swamps and usually have sulphur contents less than 1%. Both in coals and in modern peats, the sulphur is almost all organic when the total amount of sulphur is low, but substantial amounts of pyrite are present as well when total S > *ca.* 1%.

The key factor is obviously the relatively high sulphate concentrations in marine waters; sulphate is reduced to $H_2S$ by such anaerobic bacteria as *Desulfovibrio* and *Clostridium*. The distribution of the various forms of sulphur has been determined in some peat cores. The precise mechanism by which generation of $H_2S$ leads to fixation of sulphur as pyrite or in organic structures is still not clear.

In various peat cores from the mangrove swamps of Florida, organic sulphur was the predominant form in most samples, followed by pyritic.[60] The concentration of sulphate at all depths was close to that of average sea water (0.088%). Elementary sulphur and small amounts of $H_2S$ and FeS were also present. The same forms were reported in the soil of a salt marsh in the North Sea coast of Schleswig-Holstein,[61] where a linear relationship was noted between total sulphur and total organic carbon contents. It was pointed out in this study that if a salt marsh should cease to be flooded by the tide, all inorganic forms of sulphur will be oxidized to sulphuric acid and iron sulphates, and carbonate will be decomposed.

Similar results were noted in some Russian peats,[62] and a seasonal variation

[59] W. Spackman, D. W. Scholl, and W. H. Taft, 'Field Guide Book to the Environments of Coal Formation in Southern Florida', Geological Society of America, 1964; W. G. Smith, Ph.D. thesis, Sedimentary Environments and Environmental Changes in the Peat-forming Area of South Florida, The Pennsylvania State University, University Park, 1968.

[60] P. H. Given and R. N. Miller, in 'Symposium on Sulfur in Coals' ed. A. D. Cohen, Geological Society of America Special Publ. (accepted for publication, 1972).

[61] G. Brümmer, H. Grunwald, and D. Schroeder, *Z. Pflanzenernähr. Pueng. Bodenk.*, 1971, **128**, 208.

[62] K. I. Lukashev, V. A. Kovalev, A. P. Zhukhovitskaya, V. A. Generalova, and A. A. Sokolovskaya, *Litol. Polex. Iskop.*, 1972; No. 3 p. 34 (*Chem. Abs.*, 1972, **77**, 142 479); L. Ya. Kizil'shtein and L. G. Minaeva, *Khim. Tverd. Topl.*, 1972, **4**, 116.

in the distribution of sulphur forms was observed. The European workers all report a decreasing sulphate concentration with depth, whereas no such gradient was found in the Florida mangrove peats;[60] perhaps bulk flow of water at high tide through the peat into the porous limestone bedrock beneath it, a possibility indicated above, is a means of replenishing the sulphate concentration.

## 5 Organic Constituents of Wetland Peats

**Humic and Fulvic Acids.**—These ill-defined substances are found in soils, river waters, and marine sediments as well as in peats. It would be inappropriate to attempt any general account of their chemistry here, particularly in view of the numerous reviews available in the literature.[22,63–65] Some discussion of their role in the environmental chemistry of mires has already been presented. It is proposed here only to make some remarks on phenols in plants and peats and on their role in the origins of humic acids.

Flaig,[64,66] in his laboratory studies of the decay of lignin in wheat straw inoculated with known organisms, showed that the primary phenolic breakdown products are by no means all consumed by the organisms. They undergo further oxidations and the secondary products undergo condensations with themselves and with amino-acids to produce materials very like natural humic acids. That this is indeed a mechanism by which humic acids are produced in soils and peats is confirmed by both oxidative and reductive degradations,[67,68] which yield a range of phenols bearing obvious structural relationships to lignin [*e.g.* (1)—(3)], lignin being an ill-defined polymer of

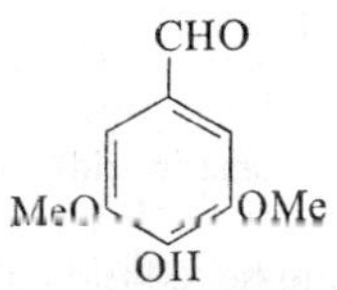


(1) syringaldehyde

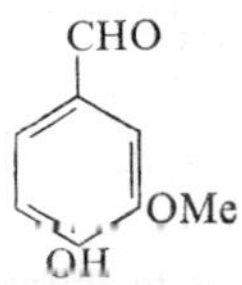


(2) vanillin

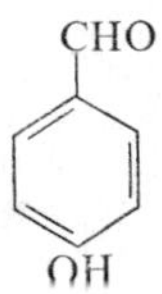


(3) *p*-hydroxybenzaldehyde

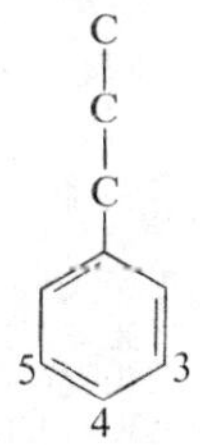


(4) phenylpropane skeleton of lignin monomer

[63] F. M. Swain, 'Non-Marine Organic Geochemistry', Cambridge Univ. Press, Cambridge, 1970.

[64] W. Flaig, in 'Advances in Organic Geochemistry 1971', ed. H. R. von Gaertner and H. Wehner, Pergamon, Oxford, 1972, p. 29.

[65] F. J. Stevenson and J. H. A. Butler, in 'Organic Geochemistry; Methods and Results,' ed. G. Eglinton and M. T. J. Murphy, Springer, Heidelberg, 1969, p. 534.

[66] W. Flaig, in 'Coal and Coal-bearing Strata', ed. D. J. Murchison and T. S. Westoll, Oliver and Boyd, Edinburgh, 1968, p. 197.

[67] C. Steelink, *J. Chem. Educ.*, 1963, **40**, 379; E. S. Lukoshko and L. V. Pigulevskaya, *Khim. Tverd. Topl.*, 1971, **3**, 39 (*Chem. Abs.*, 1971, **75**, 89 857).

[68] N. A. Burges, H. M. Hurst, and B. Walkden, *Geochim. Cosmochim. Acta*, 1964, **28**, 1547.

phenylpropane units (4) substituted in one or more of the 3-, 4-, and 5-positions, as in (1)—(3). Some derivatives of 1,3-dihydroxybenzene (resorcinol) are also found, and they are though to be derived from the B ring of flavonoids [*e.g.* (5), (6)].

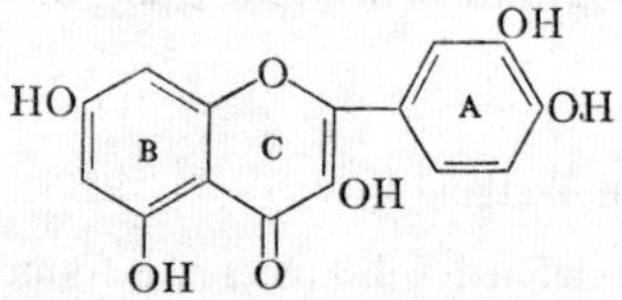


(5) quercitin, a flavonoid

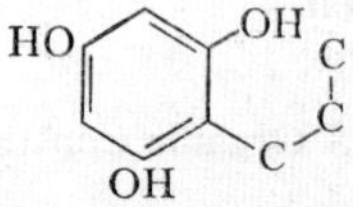


(6) scission product from B ring of flavonoid

Humic acids can also be broken down microbially.[69] In experiments on a particular soil humic acid using a *Penicillium* isolate, saligenin (7a) and salicylaldehyde (7b) were found among the products. Derivatives of saligenin are found in the leaves and bark of most species of the family Salicaceae, which includes the genera *Salix* and *Populus*. Willows and poplars had been the main elements of the vegetation populating the site where the soil was obtained before it was cleared for cultivation about 1900, so that the humic acids retained a chemical fossil in their structure for over 60 years.

Vascular plants contain a surprising variety of phenolic substances.[70] These include hydroxy- and methoxy-derivatives of cinnamic acid [*e.g.* (8) and (9)] and the corresponding cinnamyl alcohols, which are biosynthetic

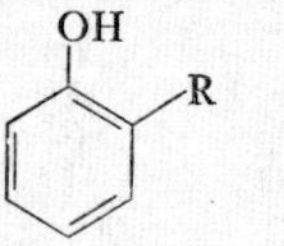


(7) a; saligenin
$R = CH_2OH$
b; salicylaldehyde
$R = CHO$

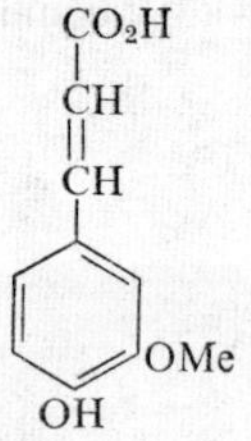


(8) ferulic acid

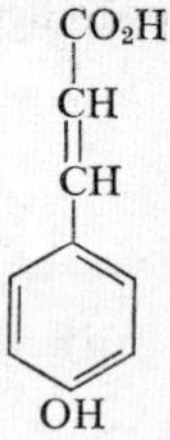


(9) *p*-coumaric acid

precursors of lignin and flavonoids. *Sphagnum* and other mosses contain many of the same phenolic substances as the vascular plants,[71] although mosses do not contain vascular tissue nor lignified cell walls. Taxonomists believe that vascular plants did not evolve from mosses, but that both had a common ancestor in green algae, their evolution representing divergent pathways. On the evidence of the phenolic constituents it would appear that a relatively

[69] S. P. Mathur and E. A. Paul, *Nature*, 1966, **212,** 646.

[70] S. M. Hopkinson, *Quart. Rev.*, 1969, **23,** 98; 'Wood Extractives', ed. A. E. Hillis, Academic Press, London, 1962, pp. 133, 159, 191, 317.

[71] H. Morita, Proceedings of the Third International Peat Congress, National Research Council of Canada, Ottawa, 1968, p. 28; G. Bendz, O. Martensson, and L. Terenius, *Acta Chem. Scand.*, 1962, **16,** 1183.

complex common ancestor, not yet identified in the fossil record, intervened between the algae and the mosses and vascular plants.

A variety of phenols of the type discussed have been identified in *Sphagnum* and other peats,[72] the concentrations varying with the seasons; the reductive degradation of fulvic acids also yields a mixture of phenols.[73] Some of the phenols identified in these studies [*e.g.* (10), (11)] do not bear a very close relationship to lignin or flavonoid structures. It seems fair to conclude that a

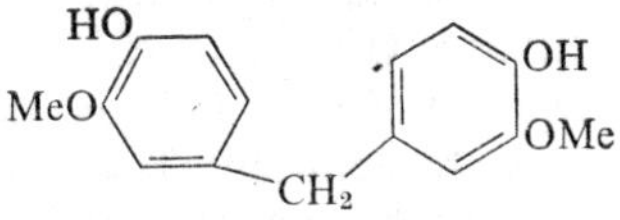


(10) diguainacylmethane

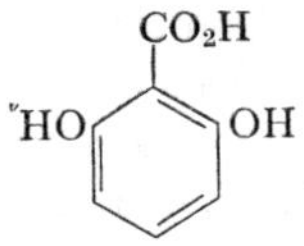


(11) α-resorcylic acid

wide variety of phenolic substances are found in peats, some deriving directly from the higher plants and some from degradations, but all may contribute to the generation of fulvic and humic acids.

A recently reported system for fractionating humic acids uses electrophoresis on polyacrylamide gel and sounds promising.[74]

**Alkanes, Fatty Acids, and Sterols.**—Two studies of the lipids of *Spartina alternifolia* in salt marshes have been reported (Johnson and Calder,[75] Jeffries[76]). Jeffries reports fatty acids only, $C_{16}$—$C_{20}$, and gives data also for *S. patens* and *Ruppia maritima*, growing near Narragansett Bay, Rhode Island. Johnson and Calder give data for hydrocarbons and fatty acids, $C_{10}$—$C_{32}$, for *S. alternifolia* and *Juncus roemerianus*, in a salt marsh near the north-west Florida coast. The relative amounts of $C_{16:0}$, $C_{18:0}$, $C_{18:1}$, and $C_{18:2}$ fatty acids are similar in the two studies of *S. alternifolia*, but whereas Jeffries finds linolenic acid ($C_{18:3}$) to account for 57% of total fatty acids, Johnson and Calder detected none. The composition of plants does vary with environment (including climate) and with the seasons, but this difference between two specimens of the same species seems surprisingly large. At the Rhode Island site, autumn frosts kill the leaves of the plants, whereas the Florida site is usually frost-free, and the plants are merely dormant in the winter.

Jeffries also studied the lipids in the muscle and digestive tract of two species of fish and a shrimp that inhabit the marsh. These were characterized by an abundance of $C_{20}$ and $C_{22}$ acids, particularly the polyunsaturated $C_{20:5}$,

72 R. I. Tsareva and T. A. Semenova, *Plastidnyi App. Zhiznedeyatel. Rast.*, 1971, p. 95 (*Chem. Abs.*, 1971, **75**, 139 728); W. Naucke, H. V. Laaser, and F. N. Tarkmann, Proceedings of the Fourth International Peat Congress, Otaniemi, Finland, 1972, vol. 4, p. 45; W. Wildenhain and G. Henseke, *Z. Chem.*, 1965, **5**, 457; *Acta Chim. Acad. Sci. Hung.*, 1972, **74**, 79; A. Morita, *Geochim. Cosmochim. Acta*, 1973, **37**, 1587.

73 W. Wildenhain and G. Henseke, *Monatsh.*, 1969, **100**, 479.

74 R. Klöcking, *J. Chromatog.*, 1973, **78**, 409.

75 R. W. Johnson and J. A. Calder, *Geochim. Cosmochim. Acta*, 1973, **37**, 1943.

76 H. P. Jeffries, *Limnol. Oceanogr.*, 1972, **17**, 433.

$C_{22:5}$, and $C_{22:6}$. He attempted to deduce from the fatty-acid distributions the relative importance of shrimp and the plant debris in the nutrition of the fish, arriving at a ratio of 1:5 approximately. It seems surprising that the possible role of algae as a food was not considered.

Johnson and Calder determined hydrocarbon and fatty-acid distributions in the sediments in which the plants were growing. The organic carbon contents of the sediments are given as around 0.1—0.5%, except in the top few centimetres, where it was 1—17% (these figures appear to be on the wet basis; water contents are not given). The changes in hydrocarbon and fatty-acid distributions followed different patterns. The hydrocarbon distribution in fresh *Spartina* and *Juncus* mostly represented cuticular wax, the alkanes with 25, 27, 29, and 31 carbon atoms being predominant. This type of distribution was perceptible in the surface sediment, but with the addition of lower alkanes, particularly $C_{17}$. At lower depths alkanes with an even number of carbon atoms became important, giving a smooth envelope peaked at $C_{23}$—$C_{24}$. Cooper[77] found quite similar phenomena in a brackish marsh populated by a sedge, *Mariscus jamaicensis*, in the Florida Everglades in which peat was accumulating. However, in a freshwater marsh dominated by the same plant (sawgrass) and in a *Nymphaea* (water-lily) marsh, the strong odd/even preference of the original plants was retained in the peats.

In the salt-marsh sediments studied by Johnson and Calder,[75] the n-$C_{16:0}$ acid remained an important constituent of the fatty acid mixtures at all levels in both the *Spartina* and *Juncus* habitats. A number of iso- and anteiso-acids in the range $C_{12}$—$C_{20}$ made their appearance in the sediments (attributed to bacterial sources), and the n-alkanoic acids in the higher range ($C_{20}$—$C_{32}$) retained a marked even/odd carbon-number preference.

Cooper's studies[77] included not only the sawgrass and water-lily habitats mentioned already but also a saline swamp populated mostly by the red mangrove, *Rhizophora mangle* L. His results were somewhat similar to those of Johnson and Calder; there were, however, some differences: (i) in some cores, particularly at lower depths, all acids below $C_{20}$, including $C_{16:0}$, represented small proportions of total fatty acids, (ii) in a number of samples, the n-$C_{20:0}$ acid was important in peat though not in the fresh plants, and (iii) homologous series of branched-chain and unsaturated acids in the range $C_{24}$—$C_{30}$ were noted in some cores at lower depth, particularly in the mangrove swamp.

Unsaturated and branched-chain fatty acids in the range $C_{12}$—$C_{28}$ have been reported in a number of other peats in different parts of the world, together with the normal saturated acids exhibiting a marked even/odd preference.[78] The longer chain fatty acids ($>C_{20}$) have been less exhaustively

[77] W. J. Cooper, 'Geochemistry of Lipid Components in Peat-forming Environments of the Florida Everglades', M.S. Thesis, The Pennsylvania State University, University Park, 1971

[78] P. I. Bel'kevich, L. A. Ivanova, and F. L. Kaganovich, *Vestsi. Akad. Navuk Belarus. S.S.R., Ser. khim. Navuk*, 1971, **5**, 111 (*Chem. Abs.*, 1972, **76**, 61 247); P. I. Bel'kevich, E. F. Dolidovich, L. A. Ivanova, F. L. Kaganovich, and Ya. V. Tserlyukevich, *Khim.*

studied in organisms, particularly algae, bacteria, and fungi, than acids of shorter chain length. It seems probable that there are sources of hydrocarbons and fatty acids in wetlands in addition to the higher plants; to identify these sources would add to our understanding of the environmental chemistry.

Various phytosterols (β-sitosterol, β-sitostanol, stigmasterol) and triterpenes (friedelin and friedelan-3β-ol) have been found in *Sphagnum* peats in Scotland and Canada as well as in a marsh peat in Israel.[79,80] These substances could certainly arise from the higher plants inhabiting the areas, but stigmasterol is also common in members of most orders of algae. A colourimetric method for determining β-sitosterol in solvent extracts of peats has been described recently.[81]

**Amino-acids.**—The occurrence of these in peats has been extensively reviewed recently,[11] and little of importance seems to have been published since the review was completed. The principal points are as follows. The concentration of free amino-acids is low, but amino-acids released by hydrolysis with boiling mineral acid can account for 2—10% of total organic matter and 30—60% of total N.[13,82] There is some reason to think that non-hydrolysable N represents amino-acids that have condensed with phenolic structures, forming N—C bonds.[66] Thus the N/C ratio in peats is considerably higher than in fresh plant material, and there is a tendency in peat ecosystems for N to be conserved or accumulated.

Amino-acids other than those common in proteins are not usually sought in sediments, but about a dozen non-protein acids were systematically determined, together with the common protein acids, in two peat cores taken in the Florida Everglades, one in a mangrove swamp and the other in a fresh water marsh populated by sawgrass, *Mariscus jamaicensis*.[11,13,83,84] The non-protein acids determined included α,ε-diaminopimelic, γ-aminobutyric, and α,γ-diaminobutyric acids, which occur in the mureide complex, the structural polymer of bacterial cell walls. These and other non-protein amino-acids associated with bacterial catabolism were commonly found in the peats, sometimes at quite appreciable concentrations. The results suggest that

79 R. Ikan, G. Stahl, and E. D. Bergmann, *Israel J. Chem.*, 1968, **6**, 485.
80 W. A. P. Black, W. J. Cornhill, and F. N. Woodward, *J. Appl. Chem.*, 1955, **5**, 494; A. J. Ives and A. N. O'Neill, *Canad. J. Chem.*, 1958, **36**, 434; R. Ikan and J. Kashman, *Israel J. Chem.*, 1963, **1**, 502.
81 E. V. Ksenofontova, M. V. Mukhiva, F. L. Kaganovich, P. I. Bel'kevich, and Ya. V. Tserlyukevich, *Vestsi Akad. Navuk Belarus. S.S.R., Ser. khim. Navuk*, 1972, **3**, 59. (*Chem. Abs.*, 1972, **77**, 91 066).
82 V. E. Rakovskii and I. A. Pal'min, *Torf. Prom.*, 1965, **42**, 23 (*Chem. Abs.*, 1965, **63**, 17 082).
83 D. J. Casagrande, 'Geochemistry of Amino-acids in Selected Florida Peats', Ph.D. thesis, The Pennsylvania State University, University Park, 1970.
84 P. H. Given, in 'Advances in Organic Geochemistry 1971', ed. H. Wehner and H. R. von Gaertner, Pergamon, Oxford, 1972, p. 69.

---

*Tverd. Topl.*, 1973, No. **1**, p. 145; M. R. Gilliland and A. J. Howard, Proceedings of the Second International Peat Congress, (Leningrad 1963), ed. R. A. Robertson, Her Majesty's Stationery Office, Edinburgh, 1968, vol. II, p. 877; J. R. Sever, T. F. Lytle, and P. Haug, *Contrib. Marine Sci. Univ. Texas*, 1972, **16**, 149.

constituents of the cells of micro-organisms may make a significant contribution to the whole organic matter of the sediment.

**Carbohydrates.**—The occurrence of these substances in peats has also been recently reviewed.[11] There are here some serious analytical difficulties; there is no satisfactory means available as yet for performing qualitative and quantitative analyses on a mixture of polysaccharides, some of which are insoluble. It has been known for many years that sugars are released by acid hydrolysis of peats, and it was assumed that these were derived from cellulose and the hemicellulose polymers of woody cell walls. Chromatographic analyses showed that although the hydrolysates contained glucose and xylose as expected, they also contained large quantities of uronic acids and appreciable amounts of arabinose, rhamnose, fucose, and ribose.[11,84,85] The analyses indicate that the extracellular slimes and capsules of micro-organisms and the carcasses of the microfauna may act as sources of polysaccharides in peats, as they do in soils.[86] However, evidence from hydrolysates is unsatisfactory and to some extent ambiguous. Lucas[51] succeeded in isolating from a peat a polysaccharide containing 76% of a uronic acid and small amounts of eight neutral sugars; its specific rotation, +31°, was consistent with α-glycosidic linkages in the polymer, in contrast to the β-glycosidic linkages in the structural polymers of the higher plants. Again we are concerned with microbial contributions to the organic sediment.

That cellulose is nevertheless preserved in peats is demonstrated by the fact that the cell walls of remanent plant tissue in peats are commonly birefringent.[84] Some recent evidence on the fate of plant polymers in peats will be dealt with below in Section 7, since it bears on some broad issues concerning the preservation of organic matter in wetland environments.

## 6 The Effect of Human Activities on Wetland Environments

The most common human activity in changing wetland environments is, of course, drainage and conversion to pasture for grazing sheep and cattle or for raising crops. The sheep of Romney Marsh were famous for the quality of their meat. This kind of activity has been practised for thousands of years and it is not proposed to discuss it here in any detail.

A little will be said, however, by way of illustration, about the Everglades of Florida, an area of which the author has some personal knowledge. An excellent non-technical account of the ecology of this fascinating area, and man's impact on it, has recently appeared.[87] The map (Figure 1) shows southern Florida as it was before man's interference. The Indian name for the

[85] P. H. Given, D. J. Casagrande, J. R. Imbalzano, and A. J. Lucas, Proceedings of the International Symposium on Hydrogeochemistry and Biogeochemistry (Tokyo 1970), D. C. Clarke Co., Washington, D.C., 1973, p. 240.
[86] U. C. Gupta, in 'Soil Biochemistry', ed. A. D. McLaren and G. H. Peterson, Marcel Dekker, New York, 1967, Ch. 4.
[87] P. Lauber, 'Everglades Country', The Viking Press, New York, 1973.

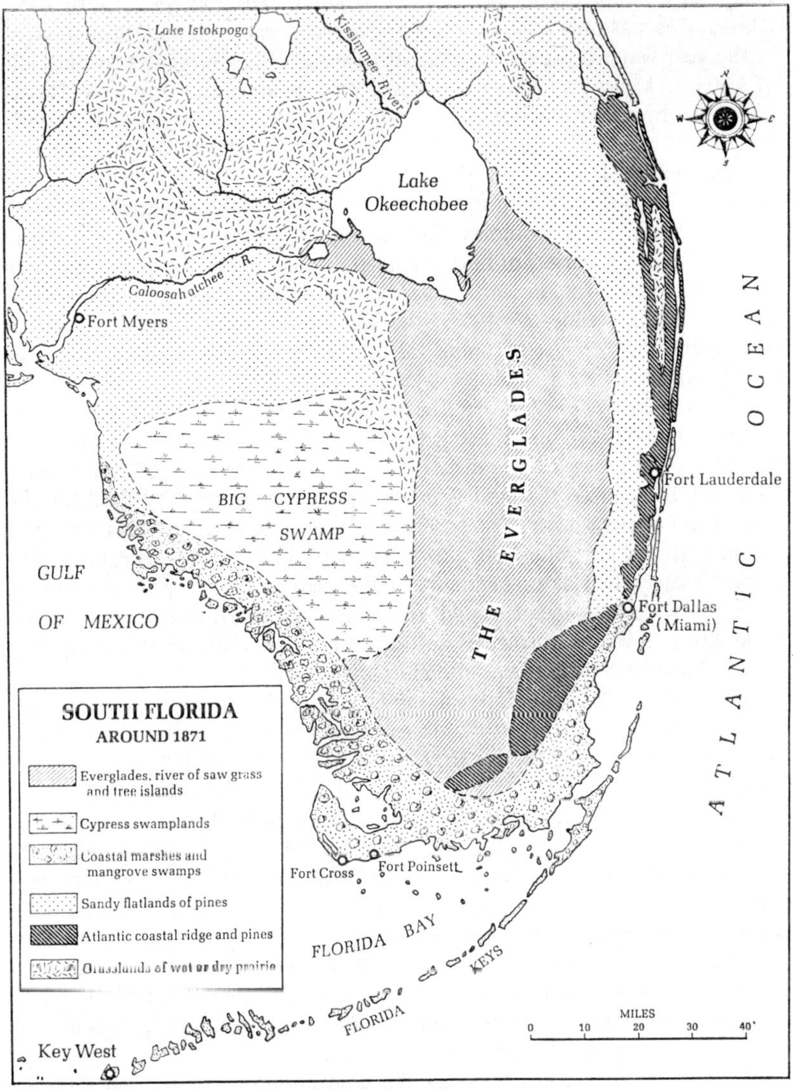


**Figure 1** *Map of Southern Florida about 1871*

(Reproduced by permission from 'Everglades Country', by P. Lauber, The Viking Press, New York, 1973. Map based on maps prepared by U.S. Army Corps of Engineers)

Everglades, Pa-hay-okee, means 'river of grass'. The southern rim of Lake Okeechobee was only seventeen feet above mean sea level, and through most of the year water flowed from the Lake through the Everglades to the Gulf of Mexico. As a river, it was very wide, very shallow, and very slow. It was populated by enormous stands of sawgrass, interspersed by water-lily ponds and tree islands. Where the bedrock is lowest, in a broad band down the centre, peat accumulated.

The marsh, together with the coastal swamps and the estuaries and bays, formed several large, complex, interacting ecosystems, which formed the habitat for a wide variety of birds, reptiles, and mammals. The basic environmental factor was the presence of very slowly flowing water over much of the area for perhaps nine months of the year. They were essentially aqueous ecosystems and their environmental chemistry (nutrition, *etc.*) was also largely aqueous.

The map in Figure 2 shows the present situation. Innumerable canals, levees (raised banks), and sluice gates have been built, which have destroyed the river of grass as a natural hydrologic system. For some 25 miles south of Lake Okeechobee the land is now drained and heavily fertilized for the cultivation of sugar cane. The function of the Conservation Areas is partly, by management of water levels, to preserve portions of the environments in approximately their natural state as wildlife refuges. But they are also part of the whole water-management system designed to prevent disastrous floods from Lake Okeechobee in hurricanes and to ensure replenishment of the aquifers from which the coastal towns and cities draw their water supply.

The National Park is unique among U.S. Parks in its ecological characteristics. Unfortunately, it is at the downstream end of the water-flow pattern, and it does not always receive the volume of water it needs to maintain the integrity of the area.

The biological effects of the destruction of the natural water-flow patterns are well known; the populations of many species of birds and of alligators have been decimated or worse since the 1930's; some species are in danger of extinction, and vegetation patterns have changed. Much less has been done on chemical aspects. When the peat is drained for agricultural use, a considerable amount of nitrogen becomes available, presumably by increased microbial activity and oxidative destruction of the organic matter (B. Volk, personal communication); also, when used for agriculture the thickness of peat decreases, probably for the same reason.

A recent study[25] has compared the concentrations of nutrients ($NH_3$, $NO_3^-$, phosphate) in the marsh waters of Conservation Area 2 and in the nearby canals. The canals receive run-off from cultivated land to the north, and contain 0.25—2.3 p.p.m. N as $NO_3^-$ and 0.06 p.p.m. N as $NH_3$, but only 0.03—0.1 p.p.m. P as phosphate. These figures are somewhat higher ($\times 2$?) than data determined at the same sites about 30 years ago. Concentrations in the surface waters of the marshes are between a tenth and a thirtieth of the values above; the figures vary with the seasons, being highest early in the

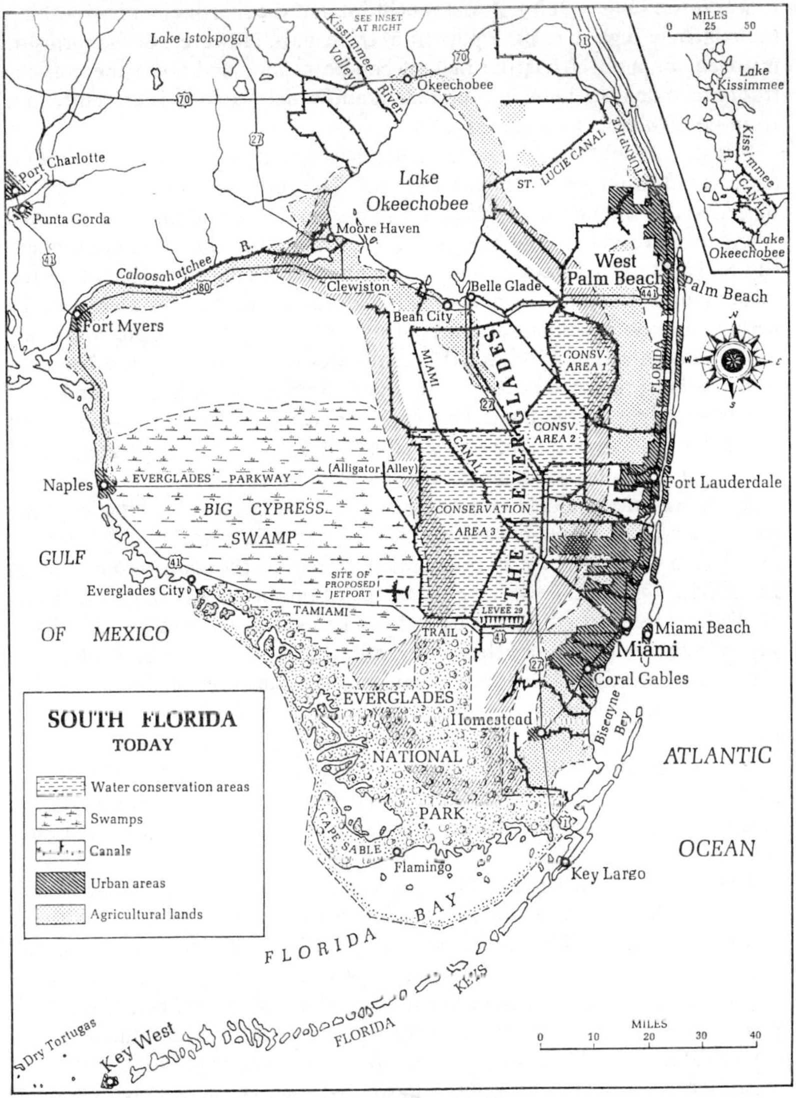


**Figure 2** *Modern map of Southern Florida*

(Reproduced by permission from 'Everglades Country', by P. Lauber, The Viking Press, New York, 1973. Map based on maps prepared by U.S. Army Corps of Engineers)

rainy season (June—July). It was concluded that the marsh biota is absorbing the relatively high nutrient inputs from the canals. However, of the forms of nitrogen, organic N has the highest concentration, and since the concentration is much the same in both marsh and canal water, it is evidently in a form not available to organisms.

It is stated that "the Everglades marsh has a purifying or 'kidney' effect on agricultural drainage discharged over it." The present author finds this an objectionably homocentric way of expressing the situation. One wonders whether algal and bacterial blooms are occurring, and what the trace element run-off is. The canal water is slightly alkaline and contains bicarbonate as the most abundant anion. The alkalinity is reduced in the marsh water, perhaps because the photosynthetic activities of algae cause $CaCO_3$ to be precipitated.[25]

The concentration of N as $NH_3$ in the freshwater areas of the National Park has been reported as 1.3 p.p.m., but phosphate is much less abundant.[88]

Fortunately, plans to build the 'jetport' shown in the map of Figure 2 were aborted in 1970; this and the associated industry, super-highways, *etc.*, would have been a terminal disaster for the southern Everglades.

A mathematical model has been constructed to represent the kinetics of the passage of DDT through the food chain in an ecosystem, and it has been tested with data obtained in a freshwater marsh at the southwest margin of Lake Erie.[89] The DDT was labelled with $^{36}Cl$ on the phenyl ring, and it was applied on an inert granular carrier by helicopter at the rate of 220 g $ha^{-1}$. Figure 3 shows the food web and a comparison of observed and calculated DDT concentrations in various organisms as a function of time. The 'fast compartment' represents transient conditions when the DDT concentration in the water was relatively high.

'Suspended material' in Figure 3 was largely phytoplankton (*i.e.* algae). A few parts in $10^9$ of DDT can appreciably reduce photosynthesis by marine algae, but it is not known whether this was a relevant factor in this freshwater marsh.

The well-known study by Blumer *et al.*[90] of the environmental effects of a small oil spill in Buzzards Bay, Massachusetts, was mostly concerned with the marine biota and sediments, but some observations were made in the coastal marshes to the north of the spill area. The concentrations of hydrocarbons in the top 2.5 cm of the marsh sediment were about 50 p.p.m. before the spill, rose to about three times this value within three months, and returned to the earlier value about 15 months after the spill. However, penetration of oil to at least 60 cm depth was detected. Though chromatographic analyses showed that bacterial degradation was occurring at depth, it was slower than at the surface.

[88] H. J. Freiberger, 'Nutrient Survey of Surface Waters in Southern Florida during a Wet and Dry Season, Sept. 1970—March 1971', U.S. Geological Survey Open File Report 72008.
[89] L. L. Eberhardt, R. L. Meeks, and T. J. Peterle, *Nature*, 1971, **230**, 60.
[90] M. Blumer and J. Sass, *Science*, 1972, **176**, 1120.

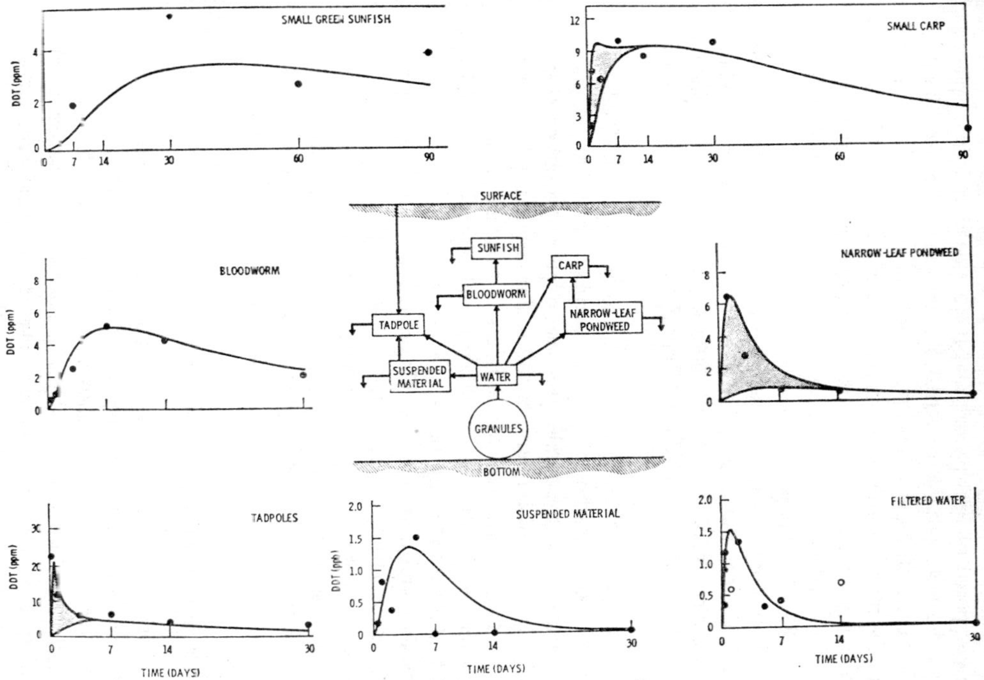


**Figure 3** *Food-chain model for selected items of the biota of a freshwater marsh. Plotted points represent observed concentrations of DDT at various times after application. Solid curves represent concentrations computed from the model equations. Shaded areas (carp, tadpole, and narrow-leaved pondweed) represent the contribution of a 'fast' compartment*

(Reproduced by permission from *Nature*, 1971, **230**, 60)

The distribution of lead has been determined in a salt marsh on the Connecticut coast, populated chiefly by *Spartina* spp.[10] The marsh is elongate and bounded on two sides by suburban housing developments. Surface samples were composed of brown ooze and matted dead and living plant material; under this were variable thicknesses of peat and estuarine silt and clay. The surface concentrations of lead were quite variable. The mean value for the part of the marsh nearest the housing development was 107 p.p.m., dropping to 82 p.p.m. at the other end of the marsh. Several storm drains discharge into the marsh from the built-up area, and concentrations of lead near these were 180—200 p.p.m.

There was a sharp drop in lead concentration with depth, it being 10—30 p.p.m. below about 30 cm; this presumably is the 'pre-automobile' level. Thus the salt-marsh sediments remove lead from inflowing waters. So far the plants and animals inhabiting the marsh do not appear to be showing unusually high concentrations of lead.

The radionuclides $^{32}P$ and $^{65}Zn$ have been used to investigate the food web in a *Spartina* salt marsh.[91] Radioactivity was determined in the surface sediment, the plant detritus, and the dominant species of micro-organisms. It was found that animals that feed on detritus and sediment and the bacteria in these concentrate the radionuclides within hours of their introduction, and peak activity is reached in a few days. No detectable uptake was found in those animals that graze the living plants.

Obviously, wetlands are most sensitive to interference with their hydrology; if this is changed too much, the ecosystem is totally altered. The organic materials of wetlands, as noted earlier, do have a remarkable ability to remove trace elements from solution. Much more work is needed to determine the extent to which toxic substances released by man's activities are in fact trapped in marshes and swamps, and whether they thereby enter the food web.

## 7 The Preservation of Organic Matter in Wetlands

Peat and organic-rich soils only accumulate in swamped environments, but not all swamped environments are peat-forming ecosystems. The author has been unable to determine the relative frequency of the two situations. In the extensive biological literature on marshes it is rare to find the organic content of the soil mentioned (it is also unfortunately not common in more chemically oriented literature to find the predominant vegetation and hydrology characterized). This is regrettable, because it makes it more difficult to seek answers to some very fundamental questions. Why does organic matter escape the carbon cycle and accumulate only in waterlogged environments? Why does it not accumulate in all such environments?

There may be several answers to the second question, applicable in different

[91] L. R. Pomeroy, E. P. Odum, R. E. Johannes, and B. Roffman, Proceedings of the Symposium on Disposal of Radioactive Wastes in Seas, Oceans, and Surface Waters, IAEA, Vienna, 1966, p. 177.

circumstances. For example, in some salt marshes, there will have been over the years a considerable input of sand. Sand will rapidly drain at each low tide, ensuring that much of the sediment is well aerated for a large fraction of each day. In these circumstances large accumulations of organic matter are improbable.

Restricted aeration is clearly a necessary condition for organic matter to accumulate in wetlands. But is it a sufficient condition? No definitive answer to this crucial question can be given at this time. Some recent work by the author and his associates suggests that the answer is negative.[92]

Filter paper was wrapped around nylon rods, which were inserted into the peat at several sites in the mangrove swamps of southern Florida. Progressively over a period of six months the cellulose was almost completely removed, starting at a point some 3 cm below the peat surface, decay proceeding downwards; paper from this depth upwards, to the top (2—3 cm above peat surface), was left intact.

$^{14}$C-labelled lignin, cellulose, and hemicelluloses were separated from tobacco leaf that had been allowed to photosynthesize in $^{14}CO_2$. At the same sites as those at which the filter paper experiments were performed, the labelled polymers were inserted into peat at depths of 12—15 cm, inside 7.5 cm diameter coring tube. After 3 months exposure, some 15% of the activity of the lignin was found in amino-acids, and most of the activity of the polysaccharides ($>$90%) had been completely removed from the system.

These various experiments show that the pure plant polymers can indeed be rapidly degraded in the natural sediment, except that there is some kind of inhibition in and above the surface layers. Yet microscopic study of thin sections of the peat shows that 20—40% of any field of view examined consists of more or less well preserved cellular tissue.[93] The cell walls of plant tissue consist of a mixture of five or six polymers, and a variety of enzymes will be needed for a concerted attack on the tissue. One would expect, therefore, that tissue would be more slowly attacked than single pure polymers. Even so there seems to be a discrepancy in rates that needs explanation.

Remaining fragments of cellulose from the filter-paper rod experiments were used as inocula for liquid media in the laboratory, fresh filter paper being provided as carbon source. Cellulolytic bacteria were observed to grow. However, a hot-water extract of the bark of the red mangrove, and also of fragments of senescent leaf, was found to inhibit growth of the cellulolytic bacteria in the liquid media. Thus the plant itself contains substances capable of inhibiting decay of its own cellulose. These can only be effective in slowing down the earlier stages of decay, since well rotted bark and leaf are by themselves inocula able to start the growth of cellulolytic bacteria.

[92] P. H. Given and C. Exarchos, Abstracts with Programs, Geological Society of America, 1973 Annual Meetings, 1973, **5** (7), p. 636; also, unpublished information.

[93] A. D. Cohen, Ph.D. thesis, 'The Petrography of Some Peats of Southern Florida (with Special Reference to the Origin of Coal)', The Pennsylvania State University, University Park, 1968.

Cellulolytic bacteria are present in the surface layers of the peat, and the reason that filter paper is not attacked in the field in these layers must be due to unsuccessful competition of cellulolytic bacteria for some essential nutrient (the filter paper is attacked when transferred to laboratory media). Some further inhibition mechanism is inferred to reduce the rate of decay of plant tissue further below the surface.

Microbiological studies[94] have shown that the populations present are quantitatively and qualitatively different in three distinct peat-forming habitats in the Florida Everglades, including those referred to above. Relatively few fragments of organic matter were found to be colonized by actively growing fungi or Actinomycetes. So far, only very small numbers of cellulolytic bacteria have been found in these studies; if this is correct, it would imply that the small population is normally almost inactive, but can be stimulated to rapid growth by the introduction of pure cellulose. Further studies of peats as distinctive habitats for micro-organisms are needed.

These results show that there are a number of zones, characterized by different sets of phenomena, in the sequence of events by which organs shed by the plants are partly transformed into peat. Moreover, there seem to be several distinct processes that reduce the rate of destruction of plant tissue. Now these investigations and conclusions refer only to one type of peat-forming environment, and it certainly cannot be assumed that the phenomena noted are universal in peat-forming ecosystems; in fact, the inhibition of cellulose decay noted in surface layers of the mangrove peat is not observed in a freshwater sawgrass site. Nevertheless, it is surely permissible to infer that in general there is a complex set of processes involved in the partial preservation of plant material in peats, and that restriction of aeration by waterlogging is not by itself a sufficient explanation of the preservation.

It is to be hoped that other investigators will address themselves to the fundamental questions posed at the beginning of this section. They will contribute to our understanding not only of basic aspects of the environmental chemistry of wetlands but also of how our vital coal reserves came to exist.

[94] C. H. Dickinson, B. Wallace, and P. H. Given, *New Phytol.*, 1974, **73**, 107.

# 4

# Environmental Organic Chemistry of Oceans, Fjords, and Anoxic Basins

BY R. J. MORRIS AND F. CULKIN

## 1 Introduction

Organic matter in the aquatic environment consists of, or is derived from, living organisms, together with a small and variable contribution from pollution. Many of the earlier investigations of the distribution of organic matter were carried out by oxidative methods which yielded no qualitative information. The determination or even simple identification of specific organic compounds, which occur at extremely low concentrations, in the presence of considerably greater amounts of inorganic matter has proved much more difficult. Nevertheless numerous organic compounds or classes of compound, which might be expected to be present from the metabolism or decomposition of organisms, have been identified, and useful summaries of the work carried out up to 1970 have been published.[1–3]

Within an aquatic environment there are a number of complex interrelating systems which must be considered for a complete understanding of the organic chemistry of the environment (Figure 1). Exchange of dissolved and particulate matter takes place at the boundaries between the water and both the atmosphere and the sediment. Organisms interact with the water (excretion and decomposition) and with one another, and bottom-living organisms interact with the deep water and with the sediments. All of these factors are, in addition, related to the particular type of aquatic environment (oceanic, inshore, anoxic, *etc.*)

This report will, therefore, be concerned firstly with recent work on the organic carbon and compounds which have been found in the waters, particulate matter, and sediments of the oceans, fjords, and anoxic basins (Figure 2) and secondly with the two important boundaries, the water–air

[1] 'Chemical Oceanography', ed. J. P. Riley and G. Skirrow, Academic Press, London Vols. 1 and 2, 1965.

[2] 'Organic Geochemistry', ed. G. Eglinton and M. T. J. Murphy, Springer-Verlag, Heidelberg, 1969.

[3] 'Introduction to Marine Chemistry', ed. J. P. Riley and R. Chester, Academic Press, London, 1971.

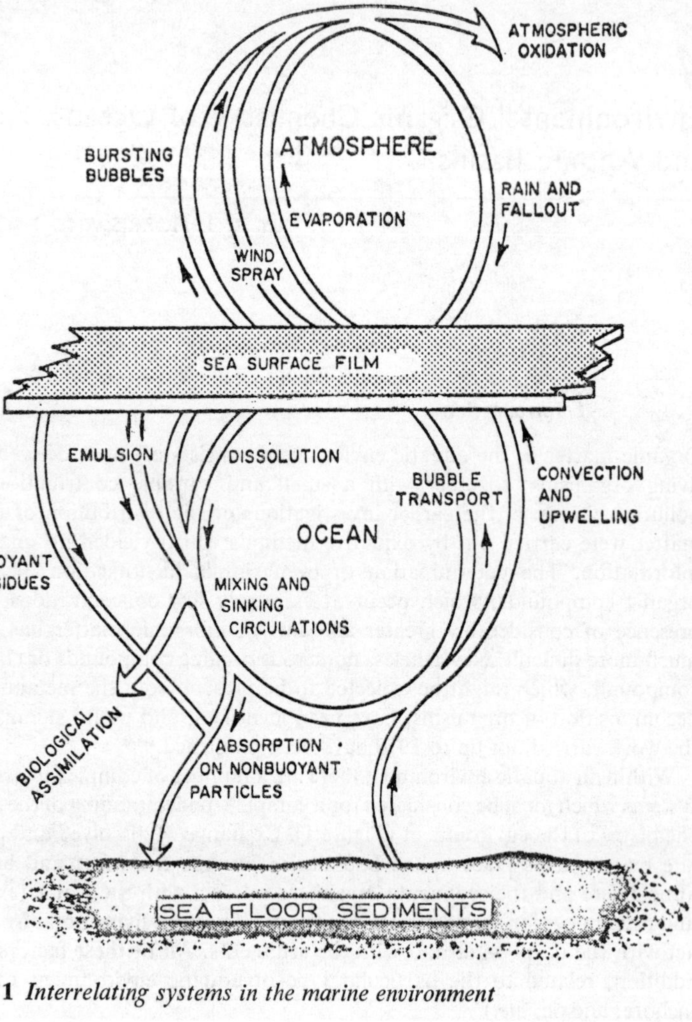


**Figure 1** *Interrelating systems in the marine environment*

and water–sediment interfaces. Discussion of the organic chemistry of living organisms (biochemistry) is considered to be outside the scope of the report.

## 2 Waters

The distinction between dissolved and particulate matter in water is an arbitrary one. It is fairly common practice to regard the dissolved fraction as that which passes through a membrane filter of pore size 0.5—1.0 $\mu$m and the particulate as that which is retained by the same filter. Concentrations of dissolved organic matter (which includes some colloidal forms) are usually less than 5 mg carbon $l^{-1}$ and exceed the particulate fraction by a factor of 10—20.

**Organic Carbon.**—Little is known of the mechanisms and processes by which dissolved (D.O.M.) and particulate organic matter (P.O.M.) are produced, maintained, or distributed. The dissolved fraction consists of the more soluble excretory and decomposition products of organisms and the particulate is mainly the more resistant fraction of the decomposition products, but the relationship between the two fractions in deep waters and the more active biochemical cycling which takes place in the upper euphotic layers is poorly understood.

The concentration of D.O.M. is generally found to be highest in oceanic surface layers,[4—6] where it has been found to be inversely proportional to the rate of primary production.[5] Thus, it was concluded that the entire biochemical cycle of organic matter in the water column, including production, decomposition, and solubilization, occurs in the upper few hundred metres. In the deeper layers the concentration of D.O.M. is fairly constant[5—7] or decreases only slightly with increasing depth.[4] It has a turnover time of *ca.* 3400 years, as indicated by carbon dating, and both the dissolved and particulate forms have been shown to be chemically and biochemically inert and appear to be of little use in the marine food-chain.[7] These findings support an earlier hypothesis concerning the nature of organic matter in the oceans.[8]

In the interstitial waters of the upper sediment layers the concentration of D.O.M. is reported to be about ten times higher than that in deep ocean waters.[4] In Baltic sediments this organic matter was found to be still labile (*i.e.* capable of further alteration), thus supporting the view that the water–sediment interface is a very important region for the diagenesis of many organic compounds.[9]

A study of the D.O.M. released during periods of primary productivity in the Western North Atlantic Ocean showed that extracellular release from living phytoplankton may account for a substantial portion of the D.O.M. and would be the main contributor to the kinetics of the D.O.M.[10] (*cf.* ref. 5).

In the English Channel, concentrations of D.O.M., P.O.M., and particulate organic nitrogen (P.O.N.) were found to increase after a spring bloom of phytoplankton, but dissolved organic phosphorus (D.O.P.) did not increase, and dissolved organic nitrogen increased only slightly.[11] In contrast, no relationship was found between total (dissolved and particulate) organic

4 N. D. Starikova, *Oceanology*, 1970, **10**, 796.
5 D. W. Menzel and J. H. Ryther, in 'Symposium on Organic Matter in Natural Waters', ed. D. W. Hood, University of Alaska, 1970, pp. 31—54.
6 A. Zsolnay, *Marine Biol.*, 1973, **21**, 13.
7 P. M. Williams, in 'Organic Compounds in Aquatic Environments', ed. S. D. Faust and J. V. Hunter, Marcel Dekker, New York, 1971, pp. 145—163.
8 A. Krogh, *Ecol. Monogr.*, 1934, **4**, 421.
9 A. Zsolnay, *Marine Chem.*, 1971, **1**, 127.
10 C. I. Choi, *Deep-Sea Res.*, 1972, **19**, 731.
11 M. W. Banoub and P. J. Le B. Williams, *J. Marine Biol. Assoc.*, 1973, **53**, 695.

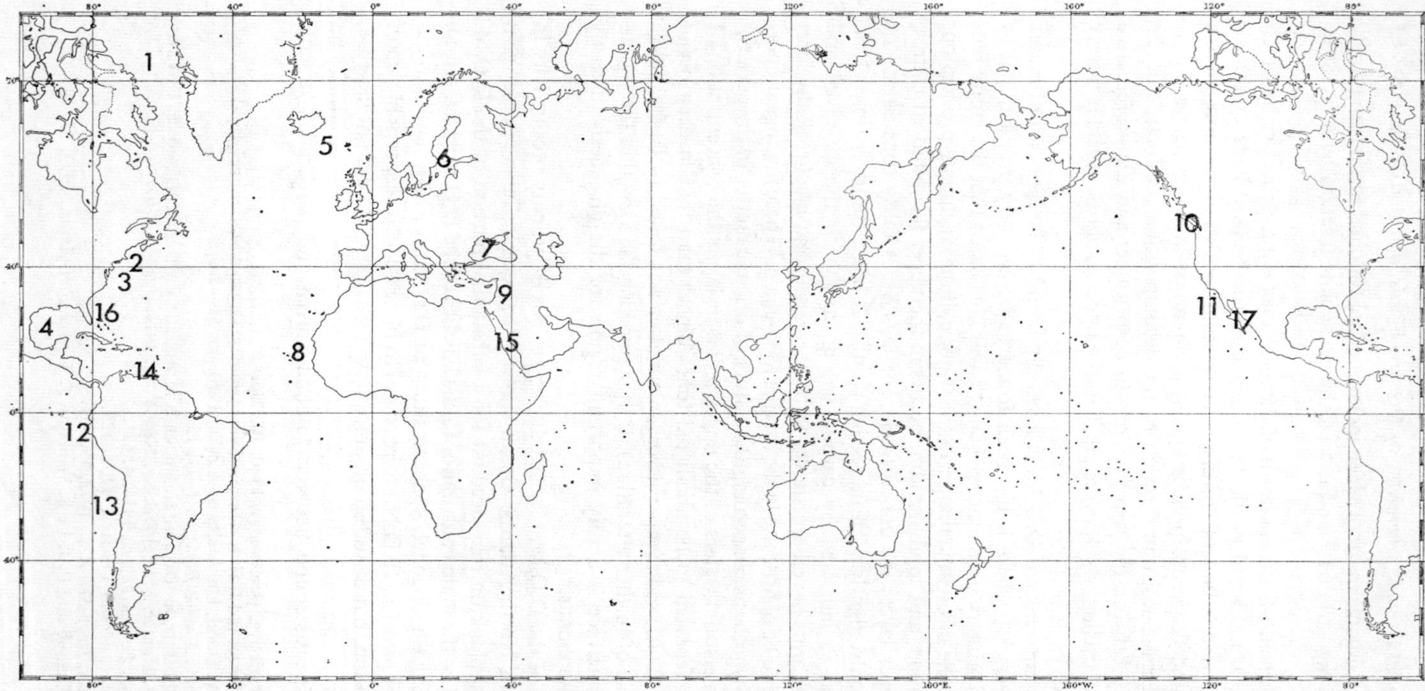


**Figure 2** *Sampling areas. Key:* 1, *Baffin Bay;* 2, *Narragansett Bay and Buzzards Bay;* 3, *Hatteras Plain;* 4, *Gulf of Mexico;* 5, *Surtsey Island;* 6, *Gotland Deep;* 7, *Black Sea;* 8, *N.W. Africa upwelling region;* 9, *Dead Sea;* 10, *Saanich Inlet and Lake Nitinat;* 11, *San Pedro Bay, Santa Barbara Basin, and San Nicholas Basin;* 12, *Peru upwelling region;* 13, *Chile upwelling region;* 14, *Cariaco Trench;* 15, *Red Sea;* 16, *Bahamas Bank;* 17, *Gulf of California*

carbon and total organic phosphorus in Baltic waters, and it was suggested that the majority of the organic matter in this region did not come from primary and secondary production but either came from mineralization of dead cells or was not of marine origin.[12]

The Amerasian Basin of the Arctic Ocean, where primary production is exceptionally low, has been found to contain higher concentrations of dissolved organic matter at all depths than do the major oceans.[13] These high D.O.M. concentrations may be due to the fact that the water masses of the Arctic have a more recent origin, relative to their surface sources, compared with other oceans, though it can also be inferred that there are different decomposition rates or transport mechanisms for particulate and dissolved organic matter in this ocean.

In the transitions from plankton, *via* suspended matter, to the sediments, the composition of organic matter is radically altered. The more readily hydrolysed compounds such as proteins and nucleic acids are lost from dead organisms, which are the principal components of suspended material. Lipids appear to be fairly stable components of primary organic matter but carbohydrates react with other compounds to form humic materials.[14]

A direct relationship between biological activity in the photosynthetic zone and the vertical distribution of carbon monoxide and gaseous hydrocarbons has been reported.[15] It was postulated that D.O.M. might be a source from which these compounds could be produced in the illuminated zone of the sea (*i.e.* the ocean could be a source of carbon monoxide).

**Lipids.**—Lipids constitute one of the more stable groups of organic compounds in animal and plant systems and they are major components of the D.O.M. of the sea. They are generally defined as those organic compounds which can be extracted with a chloroform–methanol solvent system and include such natural products as fatty acids, glycerides, wax esters, phospholipids, hydrocarbons, sterols, and sterol esters.

*Fatty Acids.* Lipids of marine organisms are generally characterized by significant amounts (5—20%) of long-chain ($C_{20}$ and $C_{22}$) polyunsaturated acids, which are not usually found in land-derived lipids. The high degree of unsaturation (up to six double bonds) of these acids makes them very susceptible to oxidation, and they have not been reported in sea water, where the dominant fatty acids are usually palmitic, stearic, and oleic (see Table 1).

Dissolved organic matter from Antarctic waters is reported to have a much higher lipid content (40—50%) than that from Gulf of Mexico waters (10—20%).[16] The difference could be due to the higher productivity of the Antarctic region and the higher lipid content of Antarctic plankton, or lower bacterial

12 S. R. Carlberg, Conference of Baltic Oceanographers, 1972, Paper No. 25.
13 P. J. Kinney, T. C. Loder, and J. Graves, *Limnol. and Oceanog.*, 1971, **16**, 132.
14 A. I. Agatova and Y. A. Bogdanov, *Oceanology*, 1972, **12**, 227.
15 D. F. Wilson, J. W. Swinnerton, and R. A. Lamontagne, *Science*, 1970, **168**, 1577.
16 L. M. Jeffrey, in 'Symposium on Organic Matter in Natural Waters', ed. D. W. Hood, University of Alaska, 1970, pp. 55—76.

**Table 1** *Typical fatty acids of the marine environment*

| | | |
|---|---|---|
| Saturated | $CH_3(CH))_{14}CO_2H$ | n-Hexadecanoic (Palmitic) |
| Monounsaturated | $CH_3(CH_2)_7CH{=}CH(CH_2)_7CO_2H$ | Octadec-9-enoic (Oleic) |
| Polyunsaturated | $CH_3(CH_2)_3(CH_2CH{=}CH)_4(CH_2)_3CO_2H$ | Eicosa-5,8,11,14-tetraenoic (Arachidonic) |
| | $CH_3(CH_2CH{=}CH)_5(CH_2)_3CO_2H$ | Eicosa-5,8,11,14,17-pentaenoic |
| | $CH_3(CH_2CH{=}CH)_6(CH_2)_2CO_2H$ | Docosa-4,7,10,13,16,19-hexaenoic |
| Branched | $CH_3\underset{\displaystyle CH_3}{\underset{\vert}{C}}H(CH_2)_{12}CO_2H$ | 14-Methylpentadecanoic (Isopalmitic) |
| | $CH_3CH_2\underset{\displaystyle CH_3}{\underset{\vert}{C}}H(CH_2)_{10}CO_2H$ | 12-Methyltetradecanoic (Anteiso-tridecanoic) |
| Isoprenoid | $CH_3\left(\underset{\displaystyle CH_3}{\underset{\vert}{C}}HCH_2CH_2CH_2\right)_3{-}\underset{\displaystyle CH_3}{\underset{\vert}{C}}HCO_2H$ | 2,6,10,14-Tetramethylpentadecanoic (Pristanic) |
| | $CH_3\left(\underset{\displaystyle CH_3}{\underset{\vert}{C}}HCH_2CH_2CH_2\right)_3{-}\underset{\displaystyle CH_3}{\underset{\vert}{C}}HCH_2CO_2H$ | 3,7,11,15-Tetramethylhexadecanoic (Phytanic) |

activity in the water column. Most of the common lipid classes were found to be present in the Gulf of Mexico waters and both saturated and unsaturated acids ($C_{14}$—$C_{20}$) were detected in solution.

In addition to $C_{11}$—$C_{19}$ saturated and unsaturated acids, $C_{12}$—$C_{19}$ iso- and anteiso-acids have been found in waters from Buzzards Bay, Massachusetts, with $C_{15}$ and $C_{17}$ being the main branched-chain acids[17] (Table 2). These branched-chain acids are common minor constituents of bacteria, marine plankton lipids, and sediments. In sea water the ratio of branched-chain to saturated acids was found to be greater than had been previously reported for sediments, but less than for marine lipids generally, suggesting a partial bacterial origin for marine iso- and anteiso-acids. The $C_{16}$, $C_{19}$, and $C_{20}$ isoprenoid acids, which are found in many marine organisms, were not detected in the sea waters.

Myristic and palmitic acids were the major fatty acids found in estuarine waters off Virginia together with unsaturated $C_{14}$, $C_{16}$, and $C_{18}$ acids. Little difference was found between the dissolved fatty acids in the estuarine and neighbouring oceanic waters.[18]

[17] M. Blumer, in 'Symposium on Organic Matter in Natural Waters,' ed. D. W. Hood, University of Alaska, 1970, pp. 153—168.

[18] T. B. Stauffer and W. G. Macintyre, *Chesapeake Science*, 1970, **11**, 216.

**Table 2** *Normal, iso-, and anteiso-acids in a coastal water sample, Buzzard's Bay, Massachusetts, Aug. 24, 1968. Values represent combined saturated and unsaturated acids as methyl esters. Total esters:* 19.7 $\mu g\ l^{-1}$ *(Concentrations/*$\mu g\ l^{-1}$*)*

| *Carbon number* | *Normal (sat. and unsat.)* | *Iso* | *Anteiso* |
|---|---|---|---|
| 11 | 0.058 | — | — |
| 12 | 0.84 | 0.0078 | 0.0056 |
| 13 | 0.089 | 0.0088 | 0.027 |
| 14 | 1.76 | 0.042 | — |
| 15 | 0.79 | 0.10 | 0.27 |
| 16 | 7.28 | 0.069 | — |
| 17 | 0.52 | 0.012 | 0.19 |
| 18 | 6.23 | 0.012 | — |
| 19 | 0.077 | <0.02 | 0.067 |
| 20 | 0.52 | — | — |
| 22 | 0.15 | — | — |

$\frac{[\text{Iso}]}{[\text{Anteiso}]}$: sea water 0.49, sediments 1.6—2, marine fats 4—8

$\frac{[\text{Normals}]}{[\text{Iso + Anteiso}]}$: sea water 22, sediments 3—7, marine fats 50—100

*Hydrocarbons.* The hydrocarbons in sea water may be derived from living organisms (biogenetic), sediments (fossil biogenetic), or pollution and are generally the stable end-products of the marine organic cycle. Some typical hydrocarbons are shown in Table 3.

Variable proportions of hydrocarbons in the D.O.M. of sea water have been reported,[19,20] principally straight-chain $C_{14}$—$C_{37}$ (maximum $C_{27}$—$C_{30}$), with

**Table 3** *Typical hydrocarbons of the marine environment.*

| | | |
|---|---|---|
| Saturated | $CH_3(CH_2)_{16}CH_3$ | n-Octadecane |
| Monounsaturated | $CH_3CH_2CH{=}CH(CH_2)_{12}CH_3$ | Heptadec-3-ene |
| Polyunsaturated | $CH_3(CH_2CH{=}CH)_5CH_2CH_2CH_3$ | Nonadeca-3,6,9,12,15-pentaene |
| | $CH_3(CH_2CH{=}CH)_6CH_2CH_3$ | Heneicosa-3,6,9,12,15,18-hexaene. |
| Branched | $CH_3CH(CH_3)(CH_2)_{13}CH_3$ | 2-Methylhexadecane |
| Isoprenoid | $CH_3\left(CH(CH_3)CH_2CH_2CH_2\right)_3{-}CH(CH_3)CH_3$ | 2,6,10,14-Tetramethylpentadecane (Pristane) |
| | $CH_3\left(CH(CH_3)CH_2CH_2CH_2\right)_3{-}CH(CH_3)CH_2CH_3$ | 2,6,10,14-Tetramethylhexadecane (Phytane) |
| | $CH_3\left(C(CH_3){=}CHCH_2CH_2\right)_3\left(CH{=}C(CH_3)CH_2CH_2\right)_2{-}CH{=}C(CH_3)CH_3$ | Squalene |

[19] G. Copin and M. Barbier, *Cahiers Oceanographiques*, 1971, **23**, 455.
[20] M. Barbier, D. Joly, A. Saliot, and D. Tourres, *Deep-Sea Res.* 1972, **20**, 305.

no odd/even chain-length preference.[17,20] The similarity of the molecular-weight pattern to that of marine algae led to the suggestion that the hydrocarbons in Atlantic and Pacific waters were derived from plankton rather than from sediment.[17,20] A similar conclusion was reached for hydrocarbons in the euphotic zone off W. Africa, where a close correlation was found between the concentrations of chlorophyll-*a* and dissolved non-aromatic hydrocarbons in the water.[21] The hydrocarbon composition was found to be fairly constant with increasing depth in the N. Atlantic and W. Mediterranean, though concentrations of total hydrocarbon were slightly lower in the deeper samples, indicating the stability of these compounds in sea water, and it was suggested that n-alkanes could be a useful geochemical indicator in the oceans.[20]

Concentrations of branched-chain hydrocarbons in sea water and in most marine organisms are usually considerably lower (by a factor of 10—100) than those of straight-chain hydrocarbons. In crude oils, however, the proportion of branched-chain hydrocarbons is much higher, and it has been suggested that they might be useful indicators of oil pollution.[17]

Short-chain hydrocarbons ($C_1$—$C_4$) have been found in waters near major ports and in the vicinity of a tanker discharging ballast water, but in water from the open ocean the only hydrocarbon detected was methane, in small amounts.[22] Longer-chain hydrocarbons from pollution sources are also found in higher concentrations in continental-shelf waters than in the open ocean. Blue-green algae and Sargassum weed have been shown to take up these hydrocarbons.[23]

The isoprenoid hydrocarbon pristane, which occurs in both phytoplankton and zooplankton, has been detected in coastal waters off the United States,[17] as has phytane, which is much less abundant in marine organisms. These two compounds, together with a higher and several lower homologues, are constituents of many crude oils, and their presence in waters has been suggested as an indicator of oil pollution.[17]

Of the olefinic hydrocarbons, squalene, which is concentrated in large quantities by sharks, and a $C_{21}$ hexaene, which also occurs in certain marine organisms,[24,25] have been detected in coastal waters.[17]

*Sterols*. Sterols always have a biogenetic origin. Although they are susceptible to autoxidation and hence would not be expected to survive unchanged for long after their release from marine organisms, there have been reports of their presence in sea water. Typical marine sterols are shown in Table 4.

The highly productive coastal waters of the Gulf of Mexico have been reported to contain cholesterol (10—17 $\mu g\,l^{-1}$), $\beta$-sitosterol (135 $\mu g\,l^{-1}$),

[21] A. Zsolnay, *Deep-Sea Res.*, 1973, **20**, 923.
[22] J. M. Brooks and W. M. Sackett, *J. Geophys. Res.*, 1973, **78**, 5248.
[23] P. L. Parker, J. K. Winters, and J. Morgan, in 'Baseline Studies of Pollutants in the Marine Environment', National Science Foundation, Office for the I.D.O., Washington, D.C., 1972, pp. 555—582.
[24] R. F. Lee and A. R. Loeblich, *Phytochemistry*, 1971, **10**, 593.
[25] M. Blumer, R. R. L. Guillard, and T. Chase, *Marine Biol.*, 1971, **8**, 183.

**Table 4** *Typical sterols of the marine environment*

*General formula*:

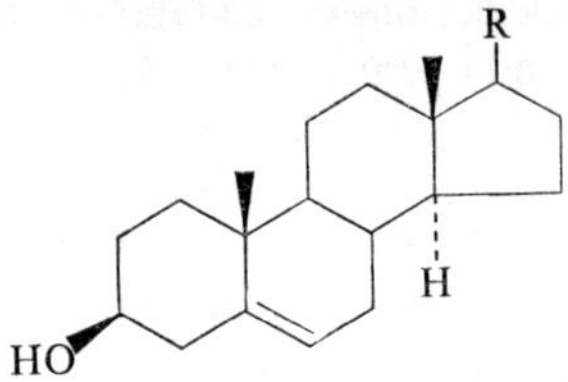


| R | *Sterol* |
|---|---|
| 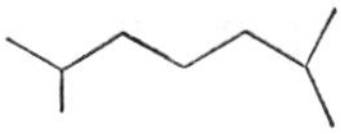 | Cholesterol |
| 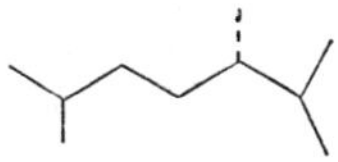 | Campesterol |
| 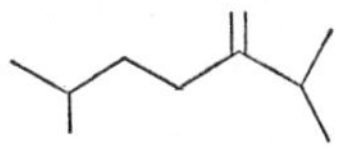 | 24-Methylenecholesterol |
| 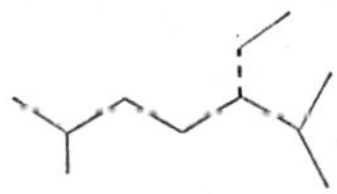 | β-Sitosterol |
| 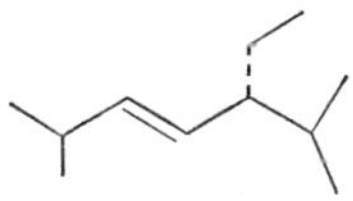 | Stigmasterol |
| 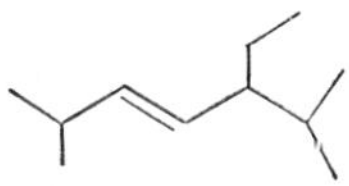 | Stigmasterol, 24-epimer |
| 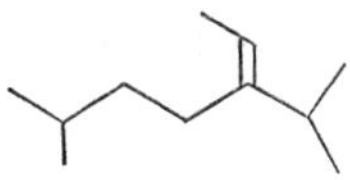 | Fucosterol |
| 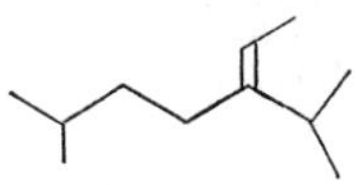 | Isofucosterol |

and stigmasterol (10—17 $\mu g\, l^{-1}$), but these values may be erroneously high because of incomplete filtration of micro-organisms from the samples.[26] In the more oceanic regions of the Gulf of Mexico no sterols were detected.[16]

Coastal and open-sea water samples from the Atlantic were found to contain similar concentration ranges of sterols (2—14 $\mu g\, l^{-1}$), possibly reflecting their concentrations in plankton and benthos.[27] Individual sterols identified were $C_{29}$ (probably $\beta$-sitosterol and its 24-epimer) and $C_{27}$ (cholesterol), which were the most abundant, $C_{28}$ (campesterol) and the diunsaturated sterols stigmasterol, fucosterol, and isofucosterol. Another sterol, tentatively identified as 24-methylenecholesterol, was found in varying concentrations, coastal waters generally being richer than near-surface open-ocean waters, while concentrations in deeper oceanic waters were very low. Sterols have also been reported in Mediterranean waters[19] and in suspended matter from Japanese coastal waters.[28]

**Amino-acids.**—Most of the protein-forming amino-acids have been detected in sea water. They are usually in the form of peptides and are derived principally from degradation of animal and plant protein. Some contribution is also likely from excretory products of organisms, since it has been shown that, under dark conditions, some species of phytoplankton metabolize protein more rapidly than carbohydrate or lipid during respiration.[29] Some of the amino-acids commonly found are listed in Table 5.

Dissolved amino-acids occur fairly uniformly throughout the water column. The most abundant are usually the neutral amino-acids, particularly metabolic breakdown products such as glycine, serine, and ornithine. Acidic and basic amino-acids are usually only present in very small amounts, or absent altogether.[29—33] Concentrations in the Irish Sea[30] and the English Channel[33] varied little throughout the year. In N. Atlantic surface waters, however, varying proportions of the free amino-acids were found and it was concluded that they resulted mainly from hydrolysis products of mixed phytoplankton populations.[32] At intermediate depths the amino-acids were thought to be products of heterotrophic activity,[32] which can result in uptake or organic matter, consuming the equivalent of 50% of the phytoplankton production.[33] Heterotrophic turnover of amino-acids and glucose in near-surface waters of the W. Mediterranean has been found to amount to 30—50% per day, detectable down to 300—500 m, but to be undetectable at 1800 m.[34]

[26] W. S. Matthews and L. L. Smith, *Lipids*, 1968, **3**, 239.
[27] A. Saliot and M. Barbier, *Deep-Sea Res.*, 1973, **20**, 1077.
[28] A. Kanazawa and S. Teshima, *J. Oceanogr. Soc. Japan*, 1971, **27**, 207.
[29] E. T. Degens, in 'Symposium on Organic Matter in Natural Waters', ed. D. W. Hood, University of Alaska, 1970 pp. 77—106.
[30] J. P. Riley and D. A. Segar, *J. Marine Biol. Assoc.*, 1970, **50**, 713.
[31] R. Pocklington, *Nature*, 1971, **230**, 374.
[32] R. Pocklington, *Analyt. Biochem.*, 1972, **45**, 403.
[33] P. Andrews and P. J. Le B. Williams, *J. Marine Biol. Assoc.*, 1971, **51**, 111.
[34] M. W. Banoub and P. J. Le B. Williams, *Deep-Sea Res.*, 1972, **19**, 433.

**Table 5** *Common amino-acids from proteins*

| | |
|---|---|
| Neutral amino-acids | |
| Glycine | $CH_2(NH_2)CO_2H$ |
| Alanine | $CH_3CH(NH_2)CO_2H$ |
| Valine | $(CH_3)_2CHCH(NH_2)CO_2H$ |
| Leucine | $(CH_3)_2CHCH_2CH(NH_2)CO_2H$ |
| Isoleucine | $CH_3CH_2CH(CH_3)CH(NH_2)CO_2H$ |

Phenylalanine

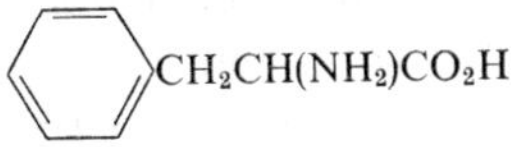


Tyrosine

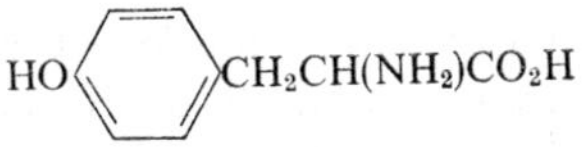


Proline

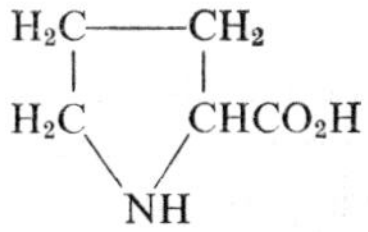


Hydroxyproline

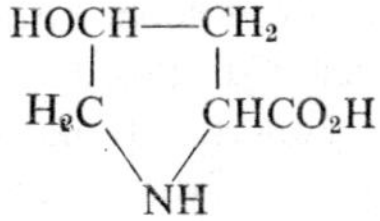


| | |
|---|---|
| Serine | $HOCH_2CH(NH_2)CO_2H$ |
| Threonine | $CH_3CH(OH)CH(NH_2)CO_2H$ |
| Cysteine | $HSCH_2CH(NH_2)CO_2H$ |
| Cystine | $[-SCH_2CH(NH_2)CO_2H]_2$ |
| Methionine | $CH_3SCH_2CH_2CH(NH_2)CO_2H$ |

Tryptophan

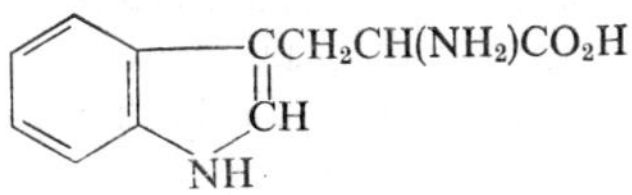


| | |
|---|---|
| Acidic amino-acids | |
| Aspartic acid | $HO_2CCH_2CH(NH_2)CO_2H$ |
| Glutamic acid | $HO_2CCH_2CH_2CH(NH_2)CO_2H$ |
| Basic amino-acids | |
| Arginine | $H_2NC(=NH)NHCH_2CH_2CH_2CH(NH_2)CO_2H$ |
| Lysine | $H_2NCH_2CH_2CH_2CH_2CH(NH_2)CO_2H$ |

Histidine

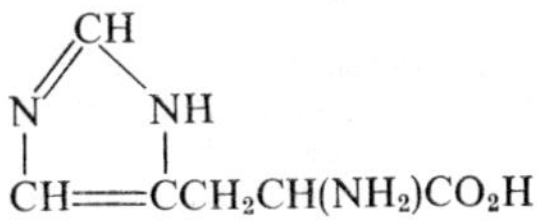

In bottom waters the amino-acids are probably the residues of processes of utilization rather than production,[32] although the higher concentrations in near-bottom coastal waters off S. California have been related to earlier findings of fairly high concentrations of free amino-acids in sediment interstitial waters, leading to the suggestion that upper sediment layers may be a major reservoir of amino-acids in the marine environment.[35]

Most of the dissolved amino-acids in sea water are combined in the form of peptides or compounds of higher molecular weight which pass through a membrane filter. These are readily hydrolysed with acid, and their amino-acid composition resembles that of the free amino-acids in having a preponderance of neutral acids.[29,30]

The amino-acid composition of particulate matter in the surface layers often suggests that it consists mostly of live phytoplankton, with varying degrees of degradation, as indicated by the presence of certain non-protein amino-acids.[29]

**Carbohydrates.**—Dissolved carbohydrates in the oceans are derived mainly from phytoplankton living in the surface layers. They occur mostly as free saccharides, in reported concentrations of 0.09—0.66 mg $l^{-1}$, and are fairly evenly distributed in most areas.[38] Pentoses and hexoses have been detected in sea water but most investigations have been concerned with the determination of glucose or total carbohydrate.

In the N. Atlantic, glucose concentrations were correlated with phytoplankton density, being highest near Bermuda and in the upwelling area off Senegal, and it was thought that microbial populations were the major users of glucose.[36] Carbohydrate in excretory products of macrophytes has been found to make a contribution to the D.O.M. which may be significant to the communities of the littoral zone.[37]

In the sea areas off Japan, dissolved carbohydrate concentrations were found to be fairly uniform in the upper 1000 m and were similar to those found in the Indian and N.W. Pacific Oceans.[38] It was suggested that particulate carbohydrate consists of at least two components, *viz.* storage polysaccharides, which degrade more slowly than protein or amino-acids in the euphotic zone, and the much more resistant structural polysaccharides which become the main components of deep-water particulate matter. This is partly supported by the findings of high carbohydrate concentrations in suspended matter from inshore waters of high productivity off Peru and Chile and much lower values in near-bottom off-shore waters.[39]

**Vitamins.**—Of the specific organic compounds which are essential for phytoplankton growth, vitamin $B_{12}$, thiamine, and biotin (Table 6) are probably

[35] M. E. Clark, G. A. Jackson, and W. J. North, *Limnol. and Oceanog.*, 1972, **17**, 749.
[36] R. F. Vaccaro, S. E. Hicks, H. W. Jannasch, and F. G. Carey, *Limnol. and Oceanog.*, 1968, **13**, 356.
[37] V. E. Erokhin, *Oceanology*, 1972, **12**, 246.
[38] N. Handa, in 'Symposium on Organic Matter in Natural Waters', ed. D. W. Hood, University of Alaska, 1970, pp. 129—152.
[39] E. A. Romankevich and I. M. Urbanovich, *Trudy Inst. Okeanolog.*, 1971, **89**, 106.

**Table 6** *Vitamins found in the marine environment*

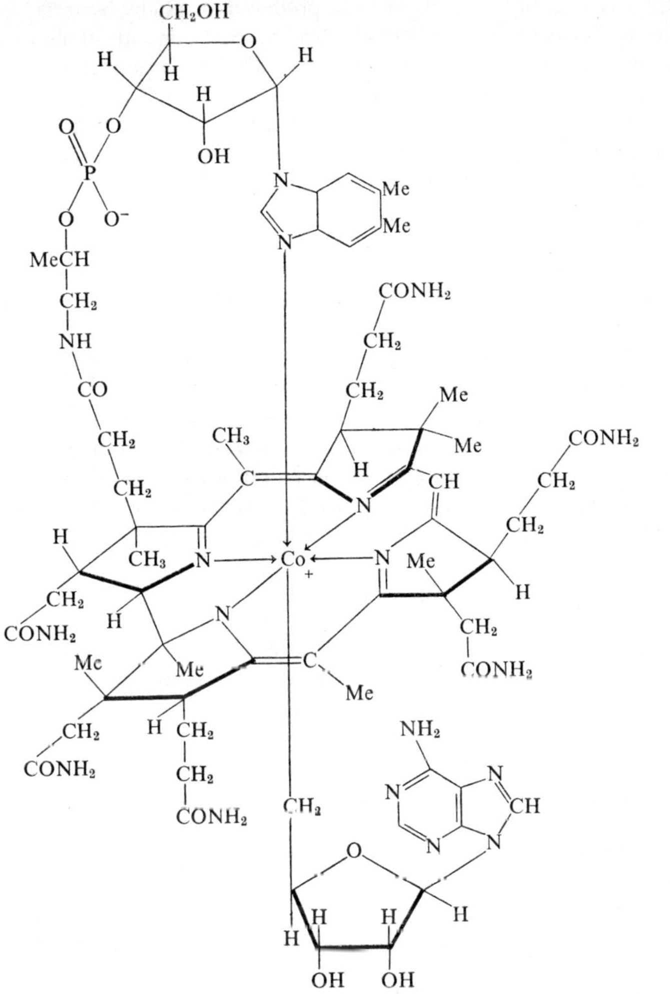

Vitamin $B_{12}$ coenzyme

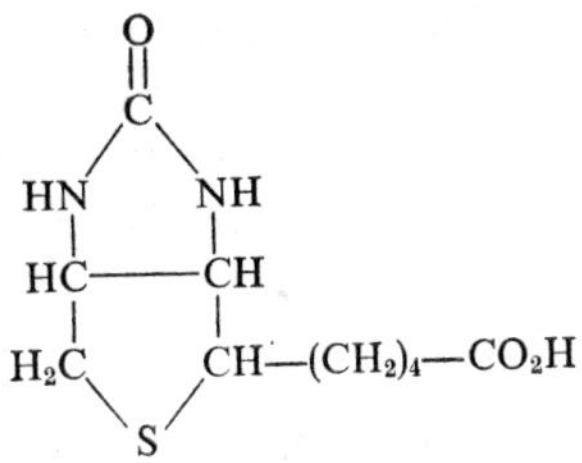

Biotin

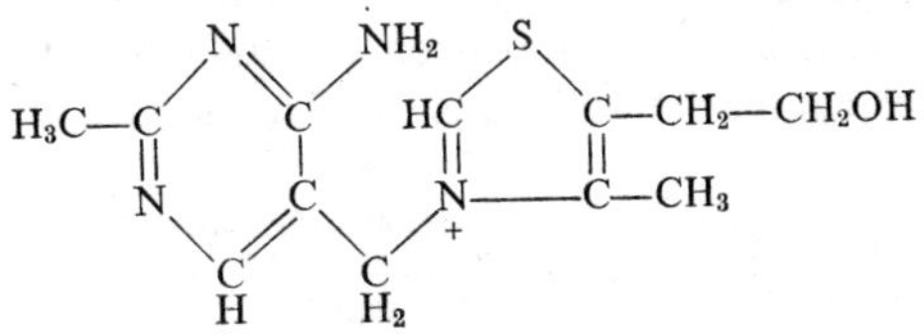

Thiamine

the most extensively studied. They are found in concentrations of a few ng $l^{-1}$ in the sea and are thought to be produced mainly by bacteria, although there are indications of other possible sources. Concentrations of chlorophyll-*a* have been found to be correlated with the concentrations of all three vitamins in coastal surface waters off Japan and with thiamine and biotin in oceanic waters.[40,41] This, together with the fact that, under laboratory conditions, some algae have been shown to produce these vitamins during growth, suggests that phytoplankton may contribute to the vitamin content of both particulate matter and water. Particulate matter from Pacific surface waters had a thiamine and biotin content corresponding to about 1% of that of the dissolved forms, whereas in coastal regions the proportions were variable but at times amounted to 144% and 54%, respectively, of the dissolved forms.[40,41]

Upwelling waters from coastal areas of the sub-Arctic Pacific Ocean have been reported to be richer in vitamin $B_{12}$ than other areas.[42] It was suggested that the source of these higher concentrations could be a deep-water maximum of vitamin $B_{12}$ concentration, such as had previously been reported in other regions.[43] In the oceanic regions, concentrations of vitamin $B_{12}$ tended to decrease slightly with increasing depth, whereas thiamine concentrations increased slightly.[42]

## 3 Sediments

The distribution of sediment types of the World Ocean is shown in Figure 3.

The organic matter in sediments is a complex mixture which includes lipids derived from phytoplankton and zooplankton metabolism and from microbiological activity on carbohydrate and protein material; original and altered protein material and its decomposition products, including amines, amino-acids, and amino-complexes; carbohydrates derived mainly from phytoplankton metabolism; pigments of various types, including chlorophyll-like compounds, carotenoids, and porphyrins; and humic materials and residual organic matter. Assessment of the relative contributions of organic matter from marine and terrigenous sources is often difficult. A number of criteria have been used, however, and these are outlined at the appropriate points in this section.

In recent years, studies of the organic geochemistry of older marine sediments have been stimulated by the Deep Sea Drilling Project of the Joint Oceanographic Institutions for Deep Earth Sampling (J.O.I.D.E.S.). Most J.O.I.D.E.S. cores, however, do not contain the upper 5—6 m of the sediment, and so comparison of studies of these cores with studies of more recent marine sediments (0—5 m) is often difficult.

[40] K. Ohwada and N. Taga, *Marine Chem.*, 1972, **1**, 61.
[41] K. Ohwada, *Marine Biol.*, 1972, **14**, 10.
[42] K. V. Natarajam, *Limnol. and Oceanog.*, 1970, **15**, 655.
[43] D. W. Menzel and J. P. Spaeth, *Limnol. and Oceanog.*, 1962, **7**, 151.

**Lipids.**—The lipids of most marine organisms consist of fatty acids (mainly esterified), alcohols, alkanes, alkenes, and sterols. Generally the hydrocarbon and sterol fractions are small (1—5% of the total lipid) compared with the fatty acid content (55—80% of the total lipid).[44] An even/odd carbon-chain preference is found in the fatty acids and alcohols and an odd/even predominance in the hydrocarbons.

During and after its incorporation into sediment, organic matter undergoes considerable alteration, and the following principal differences are found between the lipids of marine organisms and those found in sediments.

(i) The concentrations of n-alkanes in extracts from recent sediments are about ten times higher than their concentrations in the lipids of the average marine organism.[45]

(ii) The even/odd carbon-chain ratio for fatty acids is normally high in the lipids of marine organisms, but is much lower in recent marine sediments. A possible explanation of this is that the normal even-numbered fatty acids from organisms are converted in marine sediments into odd-numbered n-alkanes and fatty acids, thus:[46]

$$\underset{\text{(even number)}}{R{-}CH_2{-}COOH} \rightarrow R{-}\dot{C}H_2 + CO_2 + H^{\cdot}$$

$$R{-}\dot{C}H_2 + RH \rightarrow \underset{\text{(odd number)}}{R{-}CH_3} + R^{\cdot}$$

$$R{-}\dot{C}H_2 + (O) \rightarrow \underset{\text{(odd number)}}{R{-}COOH} + H_2O$$

(iii) The concentrations of poly-olefinic triterpenes such as squalene (an anabolic intermediate of plant and animal steroids) in marine organisms greatly exceed those in sediments. Aromatic hydrocarbons, on the other hand, which are negligible components of biolipids, are found in sediments, and their concentrations increase with depth of burial. Sterols, isoprenoids, and triterpenes have been suggested as precursors of these aromatic compounds.[45]

(iv) In marine organisms the predominantly even-numbered fatty acids occur in the chain-length range $C_{12}$—$C_{22}$ and the predominantly odd-numbered hydrocarbons in the range $C_{15}$—$C_{25}$. Higher-molecular-weight ($C_{20}$—$C_{36}$) fatty acids[47] and hydrocarbons,[47] which are not often found in the marine environment, have also been found in sediments, however, and are

44 R. J. Morris, Ph.D. Thesis, 1973, University of Southampton.
45 W. G. Meinschein, in 'Organic Geochemistry', ed. G. Eglinton and M. T. J. Murphy, Springer-Verlag, Heidelberg, 1969, 336.
46 J. E. Cooper and E. E. Bray, *Geochim. Cosmochim. Acta*, 1963, **27**, 1113.
47 S. J. Gaskell, R. J. Morris, and G. Eglinton, to be published.

**Figure 3** *Sediments of the World Ocean*

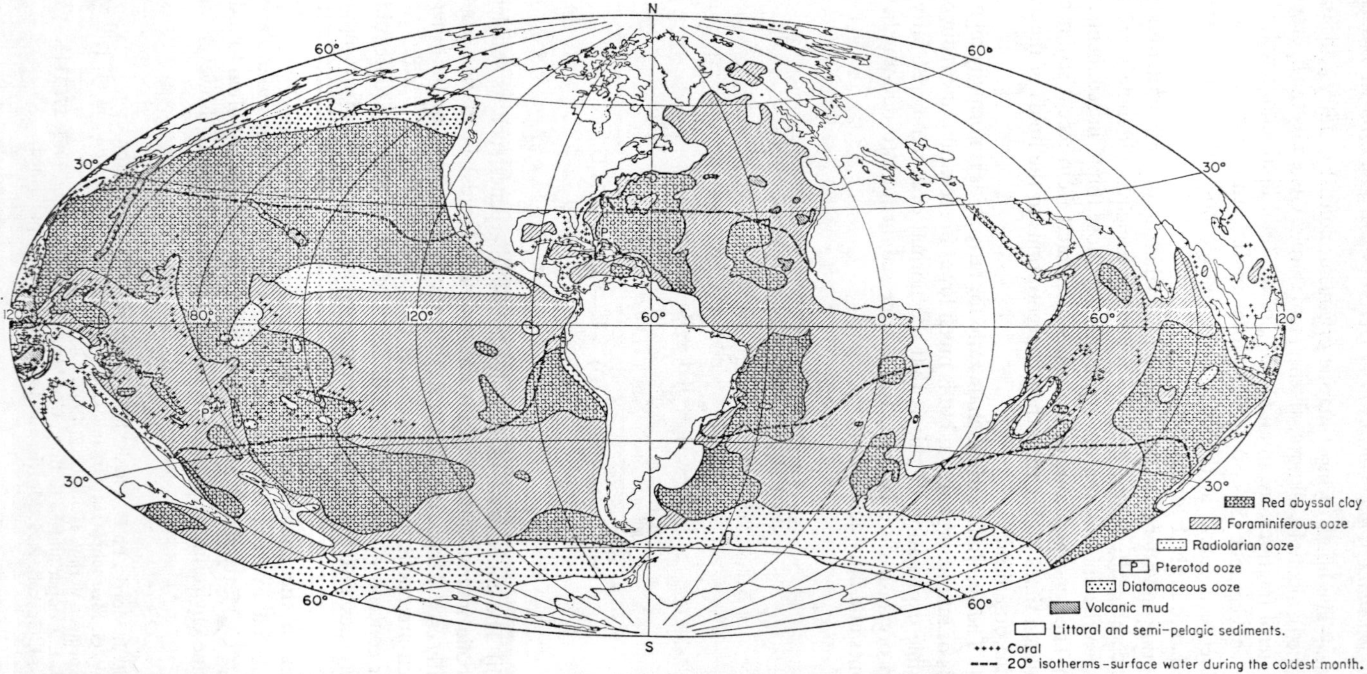


(Reproduced by permission from 'General Oceanography', ed. G. Dietrich, Wiley–Interscience, New York, 1963.)

possibly derived from terrigenous sources such as plant waxes.[48–50] The production of significant amounts of $C_{24}$—$C_{36}$ n-alkanes and $C_{23}$ and $C_{24}$ fatty acids, as well as the expected decarboxylation product, heneicosane, has been shown to occur when the saturated docosanoic acid ($C_{22}$) is heated with the mineral bentonite at 200—300 °C.[51] It seems doubtful if this type of reaction could occur in sediments, however, unless some powerful catalytic agent were present.

In recent years studies have been made of the lipids from a variety of environments. Sediments from the continental slope off Japan were found to be richer in lipids than were the red clays from the deep N.W. Pacific Ocean. In the slope sediments, polar lipids predominated, whereas the clays contained mainly weakly polar lipids, principally fatty acids and hydrocarbons. In both types of sediment, however, the content of fatty acid and hydrocarbon (especially the aromatic fraction of the hydrocarbons) increased with depth of burial.[52]

Sediments taken from a lagoon on the recently formed volcanic island of Surtsey in the N. Atlantic contained lipid with a composition similar to that of the primary organism, a green marine alga, with little indication of any contribution from higher land plants.[53]

From lipid analyses and $\delta^{13}C$ values, marine organisms are also thought to be the major sources of lipids in sediments from the Red Sea[54] and the up-welling area off N.W. Africa,[47] although in the latter case it was thought that a contribution of up to 30% may come from a terrigenous source, possibly wind-blown dust[55] or pollen.[56] The report of elemental carbon in Pacific deep-sea sediments, with evidence that it originated in continental fires,[57] is interesting in that it means that analyses of total organic carbon for sediments in this region are meaningless in terms of natural-product input.

Saturated fatty acids, usually with $C_{16}$ and $C_{18}$ predominating, have been found in sediments from the N. Atlantic (range $C_7$—$C_{21}$),[58] Baffin Bay ($C_{12}$—$C_{18}$),[59] and off N.W. Africa ($C_{12}$—$C_{28}$, maxima at $C_{16}$ and $C_{26}$, even/odd predominance).[47] A wider range of carbon-chain lengths has been reported

48 G. Eglinton, J. R. Maxwell, and R. Philp, in 'Advances in Organic Geochemistry 1973', in the press.
49 B. R. Simoneit and A. L. Burlingame, in 'Initial Reports of the Deep Sea Drilling Project', Vol. XVII, Washington (U.S. Government Printing Office), 1973, pp. 561—590.
50 B. R. Simoneit, E. S. Scott, and A. L. Burlingame, in 'Initial Reports of the Deep Sea Drilling Project', Vol. X, Washington (U.S. Government Printing Office), 1973, pp. 625—636.
51 J. W. Jurg and E. Eisma, *Science*, 1964, **144**, 1451.
52 E. A. Romankevich and S. G. Batrakov, *Geokhimiya*, 1971, **11** 1353.
53 J. R. Sever and P. Haug, *Nature*, 1971, **234**, 447.
54 J. D. Saxby, *Chem. Geol.*, 1972, **9**, 233.
55 S. R. Aston, R. Chester, L. R. Johnson, and R. C. Padgham, *Marine Geol.*, 1973, **14**, 15.
56 L. E. Florer, *Marine Geol.*, 1973, **14**, 73.
57 D. M. Smith, J. J. Griffin, and E. D. Goldberg, *Nature*, 1973, **241**, 268.
58 W. J. Cooper and M. Blumer, *Deep-Sea Res.*, 1968, **15**, 535.
59 P. L. Parker, *Contrib. to Marine Science*, 1967, **12**, 135.

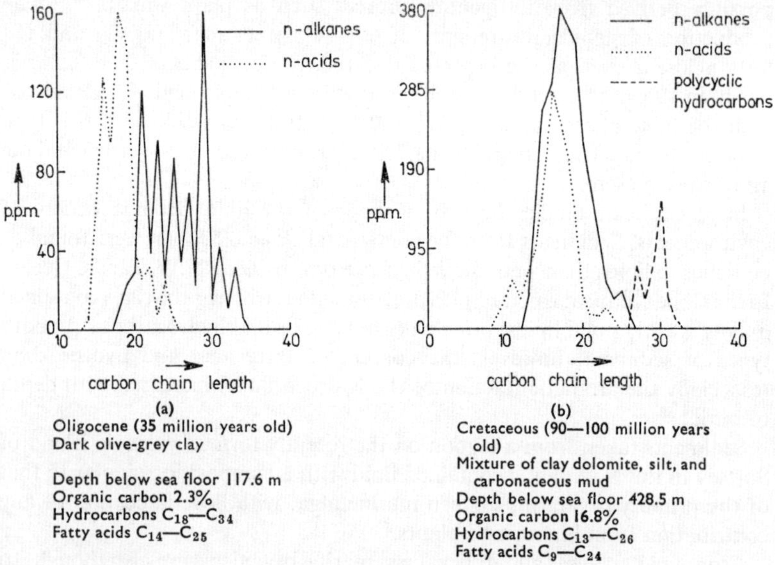


**Figure 4** *Occurrence of fatty acids and hydrocarbons in a J.O.I.D.E.S. deep-sea sediment taken off N.W. Africa (water depth* 5288 m)

for J.O.I.D.E.S. cores from the Hatteras Plain and Bahamas Bank ($C_{12}$—$C_{32}$, maxima at $C_{16}$ and $C_{24}$),[60] off N.W. Africa ($C_{14}$—$C_{25}$, as shown in Figure 4),[61] Gulf of Mexico ($C_{13}$—$C_{31}$, maxima at $C_{16}$ and $C_{24}$),[50] and ($C_{14}$—$C_{30}$, maxima at $C_{16}$, $C_{18}$, $C_{26}$, $C_{28}$),[62] and Pacific ($C_{5}$—$C_{21}$).[63]

A range of monounsaturated fatty acids, with oleic acid usually predominating, also occur widely[47,50,58—60,62,63] but more highly unsaturated acids are much less common.[47,63]

Fatty alcohols, of both odd and even number, which occur in the wax esters of the very abundant calanoid copepods[64] have been reported in sediments from Baffin Bay,[65] Gulf of Mexico,[62,65] and San Nicholas Bay[65] ($C_{12}$—$C_{30}$), together with smaller amounts of normal unsaturated alcohols.

The olive-green cells which occur in considerable quantities in the upper layers of the Atlantic Ocean constitute about 1% of the particulate matter

[60] B. R. Simoneit, E. S. Scott, W. G. Howells, and A. L. Burlingame, in 'Initial Reports of the Deep Sea Drilling Project,' Vol. XI, Washington (U.S. Government Printing Office), 1972, pp. 1013—1023.

[61] B. R. Simoneit, E. S. Scott, and A. L. Burlingame, in 'Initial Reports of the Deep Sea Drilling Project', Vol. XVI, Washington (U.S. Government Printing Office), 1973, pp. 575—600.

[62] Z. Aizenshtat, M. J. Baedecker, and I. R. Kaplan, *Geochim. Cosmochim. Acta*, 1973, **37**, 1881.

[63] B. R. Simoneit and A. L. Burlingame, in 'Advances in Organic Geochemistry 1971', ed. H. R. von Gaertner and H. Wehner, Pergamon Press, Oxford, 1972, pp. 189—228.

[64] R. F. Lee, J. C. Nevenzel, and G. A. Paffenhofer, *Science*, 1970, **167**, 1508.

[65] J. Sever and P. L. Parker, *Science*, 1969, **164**, 1052.

even at 4800 m depth[66] and, although their chemical composition is not known, they may make a contribution to the lipid content of deep-water sediments.

Some indication of the origin of hydrocarbons in recent sediments and in J.O.I.D.E.S. samples is often given by their analyses. They usually consist mainly of n-alkanes with chain-lengths varying from $C_{12}$ to $C_{50}$.[47,50,60—63,67,68] The most abundant chain-length depends on the sediment, *e.g.* $C_{29}$ and $C_{31}$ (N.W. Africa),[47] $C_{17}$ (N.W. Africa)[61], longer than $C_{24}$ (Gulf of Mexico),[50] and $C_{29}$ and $C_{31}$ (Gulf of Mexico and W. Atlantic).[62] Higher proportions of $C_{17}$, which is the typical marine lipid hydrocarbon, and an odd/even carbon-chain-length preference are both taken to indicate a marine origin. A lack of odd/even preference and the presence of hydrocarbons of high molecular weight[47,61] (unless there is evidence of chain elongation[51]), together with the presence of terpenoid hydrocarbons,[63] have been taken to indicate some terrigenous input.

Sterols are minor components of lipids in sediments and have not been very extensively reported. In Baffin Bay sediments,[69] cholesterol and campesterol were the major sterols, whereas in San Pedro Basin stigmasterol and β-sitosterol were the main ones. In neither sediment was 24-methylenecholesterol detected,[69] which is in line with a later report that the concentration of the sterol decreases with increasing depth in the water column.[27] From the high degree of structural diversity of these sterols it was suggested that sterols might give significant information about the type of biological community associated with various sediments. Similar sterol structures ($C_{26}$—$C_{29}$ skeletons) were reported in sediments from the Pacific continental shelf.[63] The 40 cm level of a Black Sea core was reported to be organically rich (18%) compared with the rest of the core. At this level the organic fraction consisted mainly of sterols and fatty acids, the principal sterol being 4-methylergostanol.[70] Apart from those mentioned above, no specific sterols have been identified recently in sediments, although sterones, steranes, sterenes, and triterpenoids have been found in Atlantic sediments[61] and terpenoid compounds have been reported in Gulf of Mexico sediments.[50]

**Amino-acids.**—Variations in total amino-acid content and in the organic carbon/organic nitrogen ratios which have been found in Argentine Basin sediments have been interpreted in terms of sedimentary environment and diagenetic changes following burial. It was suggested that fluctuations in organic carbon/organic nitrogen ratios were related to alterations in climate during Quaternary times.[71]

In J.O.I.D.E.S. cores, after hydrolysis, the ratio of free to total amino-acids

[66] R. O. Fournier, *Limnol. and Oceanog.*, 1971, **16**, 952.
[67] E. E. Bray and E. D. Evans, U.S. Geol. Survey Prof. Paper No. 581-C, 7 pp.
[68] C. B. Koons, *Geochim. Cosmochim. Acta*, 1970, **34**, 1353.
[69] D. Attaway and P. L. Parker, *Science*, 1970, **169**, 674.
[70] B. R. Simoneit, in 'The Black Sea: Its Geology, Chemistry and Biology', ed. E. T. Degens and D. A. Ross, American Society of Petroleum Geologists, 1974, p. 333.
[71] F. J. Stevenson and C. N. Cheng, *Geochim. Cosmochim. Acta*, 1972, **36**, 653.

never approached the theoretical value of 1.0 if the amino-acids became bound to kerogen-like material.[72] The ratios of glycine/alanine and aspartic acid/glutamic acid decreased with increasing age. The presence of threonine and serine in older samples was unexpected because of their thermal instability, and it was thought to indicate contamination.[72]

The protein content of zooplankton is higher than that of phytoplankton but the general amino-acid composition of both types of protein is similar.[29,73] Reports of the fate of these amino-acids when they are incorporated into sediments are not in complete agreement. The content of combined amino-acids in sediments from the Indian and Atlantic Oceans has been found to decrease rapidly in the upper 5 m and, relative to the theoretical amino-acid input from plankton, the order of abundance was found to be: neutral straight-chain (glycine and alanine) > aromatic (tyrosine and phenylalanine) > basic (histidine and arginine).[29]

In N.W. African coastal sediments, however, the order of abundance was found to be: basic straight-chain > basic heterocyclic and neutral aromatic > neutral heterocyclic, branched-chain, and sulphur-containing.

Aromatic, hydroxy, and neutral straight-chain amino-acids were thought to be more readily degraded.[73]

In Gulf of Mexico and W. Atlantic cores the acidic amino-acids were found to be the least and the neutral ones the most stable.[62] Evidence was also found of non-protein amino-acids (*e.g.* β-alanine and α-aminobutyric acid) being generated by decarboxylation of protein amino-acids within the sediments.

Amino-acids in living organisms have mainly the L stereoisomeric configuration but in the geological (sediment) environment epimerization may take place, resulting in a mixture of D- and L-enantiomers. The extent of racemization depends on the age, temperature, and matrix of the sediment so, if one variable is known for a given matrix, the other can be calculated from the experimentally determined rate of racemization. The epimerization of L-isoleucine to L-alloisoleucine has been used to estimate the age of sediments 400 000 years old[74,75] and, in conjunction with $^{14}C$ dating, as a palaeotemperature indicator.[76,77] In a study of the epimerization of seven amino-acids, glutamic acid and β-alanine were found to have the fastest rates, followed by proline, phenylalanine, valine, leucine, and isoleucine.[78]

**Carbohydrates.**—Although some carbohydrates are fairly rapidly hydrolysed, the structural polysaccharides such as cellulose are much more resistant. The only recent work on this subject concerned carbohydrates in the bottom sediments and interstitial waters from the Peru–Chile region. In the upper

[72] J. F. Wehmiller and P. E. Hare, in 'Initial Reports of the Deep Sea Drilling Project', Vol. IX, Washington (U.S. Government Printing Office), 1972, pp. 903—909.
[73] R. J. Morris, to be published.
[74] J. L. Bada, B. P. Luyendyk, and J. B. Maynard, *Science*, 1970, **170**, 730.
[75] J. Wehmiller and P. E. Hare, *Science*, 1971, **173**, 907.
[76] M. C. McKenna, J. L. Bada, and B. P. Luyendyk, *Science*, 1971, **172**, 503.
[77] J. L. Bada, R. Protsch, and R. A. Schroeder, *Nature*, 1973, **241**, 394.
[78] K. A. Kvenvolden, E. Peterson, and F. S. Brown, *Science*, 1970, **169**, 1079.

layers of sediment high values [$10^4$ $\mu$g (carbohydrate) (g sediment)$^{-1}$] were found inshore, decreasing to $10^2$ $\mu$g g$^{-1}$ in the deep off-shore areas. A similar trend was observed for interstitial waters. A correlation was found between carbohydrate content and biomass density.[39]

**Pigments.**—There are few recently published data on pigments in sediments. Low and variable concentrations of chlorophyll-*a* and tetrapyrrole pigments of the chlorin type were found to be most commonly associated with recent sediments from the Caribbean[79,81] and the Cariaco Trench.[80] Preliminary analyses suggested that the first step in diagenetic degradation of chlorophyll is loss of magnesium followed by reduction of the double bond in the phytol side-chain. In J.O.I.D.E.S. cores from the Gulf of Mexico and the W. Atlantic, concentrations of chlorins and porphyrins either decreased rapidly with increasing depth or were undetectable.[62,82]

**Humic Acids.**—Humic and fulvic acids form a major part (up to 70%) of the organic matter in sediments.[83] They are thought to be formed from carbohydrate and proteins, and there is experimental evidence that they can be formed *in situ* in the water from degradation products of plankton and are not necessarily transported from the continents.[83] Quinone compounds in the humic matter have been shown to be important in the complexing of metals in sediments and soils.[84]

**Vitamins.**—Concentrations of vitamin $B_{12}$, thiamine, and biotin in Pacific sediments have been reported to be *ca.* 1000 times higher than those reported for sea waters, with a general tendency for the concentrations to decrease slightly with increasing depth in the core.[85] These findings are significant in view of an earlier suggestion[86] that sediments could be an important source of vitamin $B_{12}$.

## 4 Anoxic Basins and Fjords

In the bottom waters of certain basins and fjords, where water circulation is very restricted, dissolved oxygen is used up in the decomposition of organic matter and is not replenished from the surface layers. A high input of organic material may also contribute to the formation of anoxic conditions. The only life which can survive in the anaerobic bottom waters and sediments is reducing bacteria. Under these conditions the degradation of organic matter would

79 E. W. Baker, in 'Initial Reports of the Deep Sea Drilling Project', Vol. IV, Washington (U.S. Government Printing Office), 1970, pp. 431—438.
80 E. W. Baker and G. D. Smith, in 'Initial Reports of the Deep Sea Drilling Project', Vol. XX, Washington (U.S. Government Printing Office), 1973, pp. 943—946.
81 J. S. Bunt, C. C. Lee, and E. Lee, *Marine Biol.*, 1972, **16**, 28.
82 F. S. Brown, M. J. Baedecker, A. Nissenbaum, and I. R. Kaplan, *Geochim. Cosmochim. Acta*, 1972, **36**, 1185.
83 A. Nissenbaum and I. R. Kaplan, *Limnol. and Oceanog.*, 1972, **17**, 570.
84 M. A. Rashid, *Chem. Geol.*, 1972, **9**, 241.
85 K. Ohwada and N. Taga, *J. Oceanogr. Soc. Japan*, 1969, **25**, 123.
86 A. F. Carlucci and S. B. Silbernagel, *Limnol. and Oceanog.*, 1966, **11**, 642.

be expected to be different from that which takes place under aerobic conditions. Qualitative and quantitative differences between the organic matter present in reducing and oxidizing regions of Lake Nitinat,[87] the Dead Sea,[88] and the Baltic Sea[89] have been reported, and there were indications that organic matter was better preserved under reducing conditions. Other investigations of organic matter in the Black Sea, Cariaco Trench, Saanich Inlet, and Lake Nitinat, however, revealed no clear evidence of enhanced preservation and led to the suggestion that the greater quantities of organic matter found in other anoxic sediments may represent faster rates of sedimentation rather than slower rates of decomposition.[90]

Concentrations of free amino-acids have been found to be similar in the surface waters and deep waters of the Black Sea, with a minimum at 150—200 m.[91] The bound amino-acids were variable in the surface layers but were fairly constant below 150 m.[91] Further investigations in the same region showed that the concentrations of (dissolved) amino-acids in the water were greater than those in the suspended matter, being highest in the upper 50 m, and decreasing at the base of the oxygenated zone then increasing again at the bottom of the $H_2S$ zone.[92] The ooze solutions were found to have a higher amino-acid content than the overlying water but the amino-acid composition of these solutions was similar to that of the sediments.[92]

In the carbohydrate fraction of the organic matter in water, suspended matter, and the surface of bottom sediments of the Black Sea, galactose, glucose, arabinose, xylose, mannose, ribose, and rhamnose have been identified. The carbohydrate content of the water and suspended matter was highest in the upper layers and lowest at 50 m and 2000 m depth, but there was no regular pattern in the vertical distribution, although the carbohydrate content appeared to be related to the total organic matter.[93]

The lipid composition of sediments from the Black Sea suggested that terrigenous detritus had almost obscured the marine components.[70] Fatty acids were mainly saturated, in the range $C_{12}$—$C_{32}$, with an even/odd preference and high molecular weight maxima, together with some mono-unsaturated acids. The lipid fraction consisted mainly of n-alkanes ($C_{15}$—$C_{33}$) with odd–even preference, together with isoprenoid hydrocarbons (phytane and pristane) and iso-alkanes. Steroid and triterpenoid compounds were also detected.[70]

The hydrocarbon fraction of deep sediments from the Dead Sea was found to contain mainly (80%) $C_{15}$ and $C_{17}$ n-alkanes, but proportions of these two

87 D. D. Adams and F. A. Richards, *Deep-Sea Res.*, 1968, **15**, 471.
88 A. Nissenbaum, M. J. Baedecker, and I. R. Kaplan, *Geochim. Cosmochim. Acta*, 1972, **36**, 709.
89 A. Zsolnay, *Kieler Meeresforsch.*, 1971, **27**, 135.
90 F. A. Richards, in 'Symposium on Organic Matter in Natural Waters', ed. D. W. Hood, University of Alaska, 1972, pp. 399—411.
91 N. D. Starikova and L. N. Korzhikova, *Oceanology*, 1969, **9**, 509.
92 N. D. Starikova and L. N. Korzhikova, *Geochem. Internat.*, 1972, **9**, 142
93 N. D. Starikova and O. G. Yoblokova, *Oceanology*, 1972, **12**, 363.

hydrocarbons were much lower (<20%) in shallow-water samples.[88] Saturated and unsaturated acids (mainly palmitic, palmitoleic, stearic, and oleic) were also found but there was no evidence of formation of hydrocarbons by decarboxylation of fatty acids.

Hydrocarbons in the water column and in the interstitial waters of shallow sediments from Saanich Inlet were found to be similar, except that the water column contained more long-chain ($>C_{28}$) alkanes and less $C_{18}$ alkanes and phytane. Interstitial waters from deeper cores contained three times more n-alkanes (mainly $>C_{22}$) than did the shallow ones, but less phytane. The fatty acid content of the water column and deep-sediment interstitial waters was low, but shallow-sediment interstitial waters were much richer, mainly in palmitic acid, which is the dominant fatty acid of plankton from the area. It was suggested that the long-chain alkanes found at depth in the sediment interstitial waters may originate from bound fatty acids or from rearrangement, addition, and decarboxylation reactions of the lower-molecular-weight fatty acids found in the upper sediment layers. The interstitial waters were found to contain amino-acids of high-molecular-weight and carbohydrates, which increased in concentration with depth, suggesting that they might act as precursors for the formation of complex polymers, humic and fulvic acids, and eventually kerogen.[94] The phytane and $C_{18}$ unsaturated hydrocarbons, which were also found in the interstitial waters, were thought to be generated by decomposition of cellular material.[94] Other analyses of Saanich Inlet sediments have shown that concentrations of amino-acids, fatty acids, fulvic acids, and chlorins decrease with increasing depth whereas those of aliphatic alkanes and humic acids show little change. There was no strong evidence of conversion of fatty acid into hydrocarbon or of *in situ* breakdown of chlorophyll to isoprenoid hydrocarbon. From $^{13}C/^{12}C$ determinations it was inferred that the humic acids had a planktonic rather than terrigenous origin.[82]

Plankton-derived olefinic hydrocarbons were found only in the upper water layers of the Gotland Deep. Below the euphotic zone only the more stable saturated hydrocarbons remained.[95]

## 5 Water–Atmosphere Interface

As well as being the interface where oil pollution is most obvious, the water surface frequently (though not always) carries a natural organic film of surface-active molecules, and it has been estimated that there is sufficient material to give almost complete coverage of the oceans by a monomolecular film.[96] This surface film has been found to contain $C_{11}$—$C_{22}$ saturated and unsaturated fatty acids, mostly esterified,[96,97] $C_{12}$—$C_{16}$ alcohols,[96,97] and high

[94] A. Nissenbaum, M. J. Baedecker, and I. R. Kaplan, in 'Advances in Organic Geochemistry 1971', ed. R. R. von Gaertner and H. Wehner, Pergamon Press, Oxford, 1972, pp. 427—440.
[95] A. Zsolnay, *Kieler Meeresforsch.*, 1971, **27**, 129.
[96] W. D. Garrett, U.S.N.R.L. Report 6201, 1964.
[97] W. D. Garrett, *Deep-Sea Res.*, 1967, **14**, 221.

concentrations of dissolved organic carbon, nitrogen, and phosphorus compounds, relative to subsurface waters.[98,99] The fatty acids in surface particulate matter from the N. Atlantic have been found to consist mainly of myristic and palmitic acids, with some long-chain polyunsaturated acids. Amounts of $C_{15}$ branched-chain acid were very low, indicating little bacterial contribution (unfortunately some of the analyses may have been affected by the presence of phytoplankton in the samples).[100]

The process of slick (film) formation is complex but it is probable that rising bubbles, convection, and diffusion bring the organic matter to the surface of the water, where a coherent monomolecular film of vertical linearly packed molecules can form under calm conditions. The longer-chain surface-active molecules tend to concentrate at the surface, and compression of the water surface by wind and currents produces a downward displacement of the shorter-chain, more hydrophilic, molecules.[96,97] Once in the surface film the organic matter can undergo alteration through a number of causes. U.v. light, for example, has been shown to produce conjugated double-bond systems in marine fatty acids which normally only contain methylene-interrupted double-bond systems. Such products are fairly readily polymerized, and the formation of insoluble particles has been observed.[101]

The presence of fatty acids in the marine atmosphere has also been reported.[102] It is presumed that this material is transported by salt particles, which are ejected by bursting bubbles or breaking waves. It can often reach the upper atmosphere. Further alteration is then possible, under the action of u.v. light and ozone, before the organic matter is returned to the sea surface in rain. The net result is the production of weathered and more durable surface films.

A number of physical and chemical properties of the interface are affected by the presence of an organic film on the surface of the water, and the following are the main ones observed or postulated:[103—105]

(a) The rate of evaporation is reduced.

(b) Localized damping of capillary waves occurs.

(c) Exchange of gases between the water and atmosphere is hindered. This can reduce the uptake of carbon dioxide by algae.[106]

(d) Light penetration is altered.

From the point of view of pollution also these surface films are important because lipophilic pollutants such as chlorinated hydrocarbons and petroleum hydrocarbons tend to become concentrated in them. There is increasing

[98] P. M. Williams, *Deep-Sea Res.*, 1967, **14**, 791.
[99] D. R. Barker and H. Zeitlin, *J. Geophys. Res.*, 1972, **77**, 5076.
[100] D. M. Schultz and J. G. Quinn, *J. Fisheries Res. Board Canada*, 1972, **29**, 1482.
[101] J. Wheeler, *J. Geophys. Res.*, 1972, **77**, 5302.
[102] W. R. Barger and W. D. Garrett, *J. Geophys. Res.*, 1970, **75**, 4561.
[103] W. D. Garrett, *J. Marine Res.*, 1967, **25**, 279.
[104] W. D. Garrett, *Deep-Sea Res.*, 1967, **14**, 661.
[105] W. D. Garrett, in 'The Changing Chemistry of the Oceans', ed. D. Dyrssen and D. Jagner, Wiley–Interscience, New York, 1972, pp. 75—91.
[106] W. Schramm, *Marine Biol.*, 1972, **14**, 189.

evidence that a number of heavy metals (*e.g.* Pb, Fe, Ni, Cu, Zn, Cd) are also concentrated in the organic and particulate matter of the surface microlayer, by a factor of $10^3$—$10^4$ over their concentrations in bulk sea water.[99,107—109] This enrichment of heavy metals in the surface film is thought to be due to complexing by the surface-active organic matter, although no specific complexes have yet been identified.

Enrichment of a number of trace elements in marine aerosols and rain has also been reported,[99,110—112] and mechanisms to account for the transfer of organic matter and associated trace elements into the atmosphere, involving enrichment on sea-salt aerosols, formed by bursting bubbles and breaking waves, have been put forward.[108,111—114]

## 6 Water–Sediment Interface

The organic matter which is deposited in the sediments is mostly the more resistant fraction which survives degradation in the upper layers of the water column. It is incorporated into the sediments in a number of ways. Planktonic and terrigenous detritus may be deposited directly in the sediment, where it will be subject to chemical or bacterial activity and become part of the pool of organic matter which is available for *de novo* synthesis by micro-organisms and larger benthic organisms. That which is not readily attacked by the microbial population may be incorporated into complex humic materials. Some organic matter is deposited in the form of complex polymers but polymerization can also take place by epitaxial growth on mineral surfaces, after adsorption from solution. In general, degradation of organic matter is favoured by oxidizing conditions and high microbial activity while preservation and polymerization are favoured by reducing conditions, adsorption on clay minerals or carbonate shells, and low microbial activity.

**Microbial Activity.**—The distribution of bacteria in the water column is uneven, as they are abundant only in the surface waters and the organic detritus. In the sediments their development is limited by the availability of assimilable organic matter, and microbial activity often occurs only in the upper few centimetres. Below 20 cm depth in the Pacific Ocean and Black Sea sediments, for example, sterile conditions have been found.[115]

107 R. A. Duce, J. G. Quinn, C. E. Olney, S. R. Piotrowicz, B. J. Ray, and T. L. Wode, *Science*, 1972, **176**, 161.
108 S. R. Piotrowicz, B. J. Ray, G. L. Hoffman, and R. A. Duce, *J. Geophys. Res.*, 1972, **77**, 5243.
109 K. H. Szekielda, S. L. Kupferman, V. Klemas, and D. F. Polis, *J. Geophys. Res.*, 1972, **77**, 5278.
110 D. H. Peirson, P. A. Cawse, L. Salmon, and R. S. Cambray, *Nature*, 1973, **241**, 252.
111 M. Komabayasi, *J. Met. Soc. Japan*, 1962, Ser. II, **40**, 26.
112 R. Chesselet, J. Morelli, and P. Buat-Menard, *J. Geophys. Res.*, 1972, **77**, 5116.
113 R. A. Baier, *J. Geophys. Res.*, 1972, **77**, 5062.
114 J. G. Quinn and T. L. Wade, in 'Baseline Studies of Pollutants in the Marine Environment', National Science Foundation, Office for the I.D.O., Washington D.C., 1972, pp. 633—663.
115 Y. I. Sorokin, in 'The Changing Chemistry of the Oceans', ed. D. Dyrssen and D. Jagner, Wiley–Interscience, New York, 1972, pp. 189—204.

The mechanism of microbial alteration of lipids has been studied by injecting doubly labelled oleic acid into an estuarine sediment (*in situ*).[116] Saturated fatty acids ($C_{12}$—$C_{18}$) were produced in about 2% yield and two possible biochemical pathways were postulated for their formation (Figure 5). One involved reduction of the double bond and removal of $C_2$ units (degradation) and the other involved $\beta$-oxidation to acetyl-CoA units followed by re-synthesis. Labelled branched-chain $C_{15}$ and $C_{17}$ saturated acids were

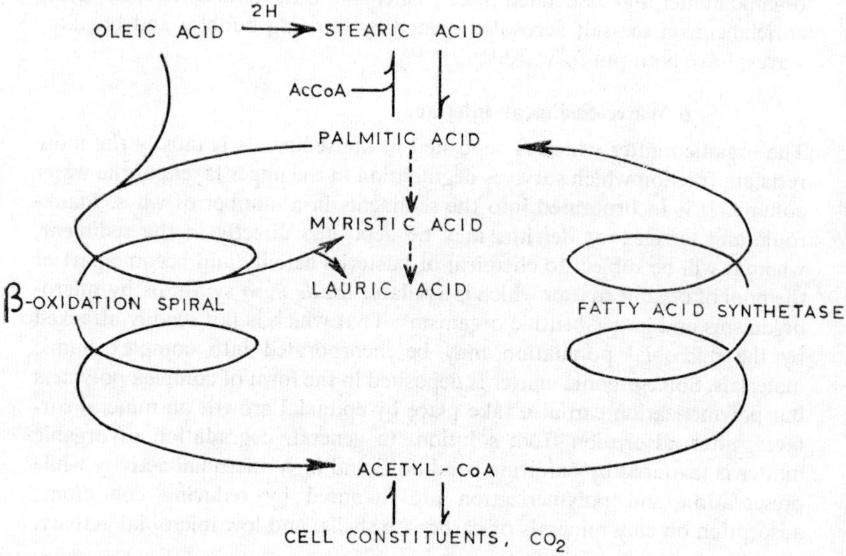


**Figure 5** *An outline of the possible pathways of oleic acid diagenesis* (Reproduced by permission from 'Advances in Organic Geochemistry 1971', ed. H. R. von Gaertner and H. Wehner, Pergamon Press, Oxford, 1972, p. 327)

also found in 0.2% yield, presumably synthesized by bacteria from acetyl-CoA. These branched-chain acids are found in only trace amounts in most marine organisms but they are major fatty acids of bacteria. Their presence in deep oceanic sediments collected off N.W. Africa (7—11% of the total saturated acids),[47] in marsh grass detritus (but not living marsh grass) from Narragansett Bay,[117] and salt-marsh sediment from N.W. Florida[118] has led to the suggestion that they can be sensitive indicators of microbial activity in depositional environments. It should, however, be recognised that these

[116] M. M. Rhead, G. Eglinton, G. H. Draffan, and P. J. England, *Nature*, 1971, **232**, 327.
[117] D. M. Schultz and J. G. Quinn, *Estuarine and Coastal Marine Sci.*, 1973, **1**, 177.
[118] R. W. Johnson and J. A. Calder, *Geochim. Cosmochim. Acta*, 1973, **37**, 1943.

branched-chain acids have not necessarily been synthesized in the sediment and that they could be part of a terrigenous input.

Rapid cycling of fatty acids, *via* a carbon pool, shortly after deposition has also been found to take place in surface sediments from Narragansett Bay,[119] where glycerides were found to be absent, presumably because of the lipolytic activity which was present. Total fatty acid content of sediments from the same area has been shown to decrease with increasing depth, and the saturated fatty acids appeared to be more stable than the unsaturated.[120]

In contrast to fatty acids, fuel-oil hydrocarbons have been shown to be much more resistant to bacterial breakdown and were still present just below the sediment surface two years after being trapped.[121] A study of the distribution of non-aromatic hydrocarbons in Central Baltic waters showed that higher concentrations were most frequently found in waters from basins which were often anaerobic or which had been in contact with the sediment surface. Because of the poor exchange between deep and surface waters in this region it was considered unlikely that these hydrocarbons were due to surface pollution but rather that they were the product of *in situ* synthesis at the water–sediment interface.[9]

The amino-acid content of sediments is also dependent on whether the environment is aerobic or anaerobic. In reducing sediments the free amino-acid content tends to be high, especially if microbial activity is low.[29,91] In an oxidizing environment the amount and nature of the free amino-acids present are closely related to microbial activity, and *de novo* synthesis of proteins from detritus and dissolved organic matter appears to be brought about by micro-organisms and burrowing animals (*e.g.* in sediments and interstitial waters of Santa Barbara Basin).[91]

Deep-water sediments from the Gulf of California have been reported to be richer in carbohydrates than were salt-marsh and shallow-water sediments from the same region. The preservation was attributed to inhibition of microbial activity by hydrostatic pressure and low-energy reducing conditions.[122]

**Sediment–Soluble Organic Compound Associations.**—Some classes of organic compound exhibit affinities for specific mineral fractions[123,124] and it has been suggested that the occurrence of some organic matter in marine sediments has resulted from its association with settling mineral particles.[125]

Fatty acids are thought to be bound to clay minerals by weak van der Waals forces and hydrogen bonds (heat of adsorption *ca.* 14.6 kcal $mol^{-1}$).[124] Adsorption from water of salinity 4‰ was found to be 3—4 times greater

119 J. W. Farrington and J. G. Quinn, *Geochim. Cosmochim. Acta*, 1973, **37**, 259.
120 J. W. Farrington and J. G. Quinn, *Nature Phys. Sci.*, 1971, **230**, 67.
121 M. Blumer and J. Sass, Woods Hole Oceanographic Institution Report 72-19, 1972.
122 F. M. Swain and J. M. Bratt, in 'Advances in Organic Geochemistry 1971', ed. H. R. von Gaertner and H. Wehner, Pergamon Press, Oxford, 1972, pp. 415—425.
123 Y. A. Romankevich and V. E. Artem'yev, *Oceanology*, 1969, **9**, 644.
124 P. A. Meyers and J. G. Quinn, *Geochim. Cosmochim. Acta*, 1973, **37**, 1745.
125 P. A. Meyers and J. G. Quinn, *Geochim. Cosmochim. Acta*, 1971, **35**, 628.

than from pure water but higher salinities had little effect. Increase in temperature (0—50 °C) and pH (6.0—8.5) both tended to reduce adsorption, however. The following order of adsorptive ability was reported for minerals:

Bentonite > Kaolinite > Illite > Montmorillonite > Calcite.[124]

Fuel oil was similarly found to be bound to mineral particles by weak physical bonds but was not easily removed by washing with saline solution.[126]

Carbonate minerals have been found to have coatings of organic matter, principally fatty acids,[127] lipids, and amino-acids.[128] The effect of these coatings is to interfere with the carbonate equilibria in sea water. It has also been found that, when heated with calcium carbonate (calcite) at 250 °C for periods of up to 300 hours, fatty acids are degraded to alkanes, possibly *via* $\beta$-cleavage of the acid.[129] Thus the formation of alkanes in carbonate rocks may be thermocatalysed by the carbonate rock itself.

As well as acting as a carrier of organic material into the sediments, clay minerals can also act as a site for further chemical reaction. In the presence of kaolin, for example, polysaccharides have been found to form from paraldehyde in solution, and fatty acids in solution can be esterified to form glycerides.[130] The asymmetric synthesis of peptides, containing mainly L-form amino-acids, from a solution of amino-acids containing kaolin has also been reported.[29]

126 P. A. Meyers and J. G. Quinn, *Nature*, 1973, **244**, 23.
127 P. A. Meyers and J. G. Quinn, *Limnol. and Oceanog.*, 1971, **16**, 992.
128 E. Suess, *Geochim. Cosmochim. Acta*, 1970, **84**, 157.
129 A. Shimoyama and W. D. Johns, *Geochim. Cosmochim. Acta*, 1972, **36**, 87.
130 G. R. Harvey, K. Mopper, and E. T. Degens, *Chem. Geol.*, 1972, **9**, 79.

# 5
# Hydrocarbons in the Marine Environment*

BY J. W. FARRINGTON AND P. A. MEYER

## 1 Introduction

The importance of the study of hydrocarbons found in the marine environment has been set forth by several authors.[1–4] Hydrocarbons in sediments have received considerable attention because of the importance of understanding the origin of crude oil. Research concerned with hydrocarbons in the marine environment has intensified during the past few years because of a substantial commitment towards understanding oil pollution problems. Published scientific literature dealing with this topic has rapidly expanded and has been accompanied by a proliferation of other publications such as technical reports and government documents. This Report focuses on the literature of the period 1970—1973 and refers to other reviews where possible.

This Report is concerned primarily with the hydrocarbons with 10 or more carbon atoms. We have not reviewed the literature dealing with low molecular weight hydrocarbons, *i.e.* methane to decane. Analytical techniques and the relative importance of processes governing the input and distribution of low molecular weight hydrocarbons compared with higher molecular weight hydrocarbons are sufficiently different to warrant the separate treatment we have set forth here for the higher molecular weight hydrocarbons. Much of our discussion concentrates on the methods of obtaining the data because an understanding of the distribution and fate of hydrocarbons in the marine environment is very much dependent on the validity and usefulness of these methods.

While omissions are inevitable, we think we have discussed the major advances in research dealing with this topic.

* This constitutes contribution No. 3310 of the Woods Hole Oceanographic Institution and contribution No. 219 of the Department of Atmospheric and Oceanic Sciences, University of Michigan.

1 W. G. Meinschein, in 'Organic Geochemistry: Methods and Results', ed. G. Eglinton and M. T. J. Murphy, Springer-Verlag, New York, 1969, p. 330.
2 M. Blumer in 'Organic Matter in Natural Waters', ed. D. W. Hood, Institute of Marine Science, University of Alaska Publication No. 1, 1970, p. 153.
3 J. R. Maxwell, C. T. Pillinger, and G. Eglinton, *Quart. Rev.*, 1971, **25**, 571.
4 M. Calvin, 'Chemical Evolution', Oxford University Press, New York, 1969.

## 2 Origin of Hydrocarbons

Hydrocarbons enter the marine environment *via* three general processes.

**Biosynthesis.**—Marine and land organisms biosynthesize hydrocarbons either *de novo* or from ingested precursor compounds. These hydrocarbons may be released during metabolism or upon the death and decomposition of the organisms. Estimates of the rate of biosynthesis of hydrocarbons by marine primary productivity are generally given as 1—10 million metric tons per year.[5–7] Estimates of the total input to the marine environment by land organisms are not available. However, substantial quantities of hydrocarbons are released from land plants to the atmosphere[8,9] where they are subject to deposition on land and on sea, and to photochemical and chemical alteration. The actual amount reaching the world's oceans is unknown.

**Geochemical Processes.**—Submarine and coastal land oil-seeps release petroleum hydrocarbons to the marine environment. The annual input rate is variously estimated at between less than 0.1 million metric tons[10] and 10 million metric tons.[11] A recent review of this subject has arrived at an annual input rate of 0.6 million metric tons per year.[12]

Weathering of soils and sediments and transport of some of the hydrocarbons in these sediments to the marine environment should also be considered as an input, although probably small relative to other sources because of slow degradation of the hydrocarbons during the weathering process. No estimates of the annual rate of input are available.

There are chemical synthesis processes which are sources of hydrocarbons. Forest fires inject an estimated 6 million metric tons of hydrocarbons per year[13] into the atmosphere. An unknown portion is eventually delivered to the marine environment. There are also chemical reactions occurring during the diagenesis of organic matter in sediments which yield hydrocarbons.

5 'The Water's Edge, Critical Problems of the Coastal Zone', ed. B. H. Ketchum, M.I.T. Press, Cambridge, Mass., 1972, p. 161.
6 D. K. Button, in 'Impingement of Man on the Oceans', ed. D. W. Hood, Wiley Interscience, New York, 1971, p. 422.
7 Calculations by the authors, based on primary productivity, average hydrocarbon content of phytoplankton, and assuming no recycling of hydrocarbons from primary productivity.
8 R. A. Rasmussen, *Environ. Sci. Technol.*, 1970, **4**, 667; F. W. Went, *Proc. Nat. Acad. Sci. U.S.A.*, 1960, **46**, 212.
9 J. L. Laster and R. Valle, *Environ. Sci. Technol.*, 1971, **5**, 631.
10 M. Blumer, *Science*, 1972, **176**, 1257.
11 R. D. Wilson, 'Background Papers for a Workshop on Inputs, Fates, and Effects of Petroleum in the Marine Environment', May, 1973, Vol. I, Ocean Affairs Board, U.S. National Academy of Sciences, Washington, D.C., p. 59.
12 'Inputs, Fate, and Effects of Petroleum in the Marine Environment', A Report of the Ocean Affairs Board, National Academy of Sciences, Washington, D.C., 1974, in the press.
13 'Background Papers for a Workshop on Inputs, Fates, and Effects of Petroleum in the Marine Environment', May, 1973, Vol. I, Ocean Affairs Board, U.S. National Academy of Sciences, Washington, D.C.

These hydrocarbons may eventually find their way into the present marine environment either by submarine exposure of sediments or by diffusion out of the sediments. No estimates are available as to rates of input. However, the annual rate is probably small.

**Anthropogenic Inputs.**—The inputs of hydrocarbons from accidental and intentional discharge of fossil fuel to the marine environment and the input of hydrocarbons from the atmosphere have been estimated and recently reviewed in detail.[12,13] The annual input rate is of the order of 1—10 million metric tons. The uncertainties of several of the estimates should be emphasized as well as the fact that these are global averages and may be quite different for any given location and any given point in time.

## 3 Analysis of Petroleum Hydrocarbons and Biosynthesized Hydrocarbons

One of the critical problems for analysts of hydrocarbons in marine samples is the detection of petroleum hydrocarbons in the presence of recently biosynthesized or native hydrocarbons. The methods of analysis for bulk oil samples and slicks resulting from accidental spills have been discussed and reviewed by other authors.[14] Techniques for analysis of recently biosynthesized hydrocarbons in marine organisms from the viewpoint of elucidation of chemical structure have been reviewed.[3,14]

**Comparison of the Composition of Petroleum Hydrocarbons and Biosynthesized Hydrocarbons.** The detection of petroleum hydrocarbons in many marine samples is complicated for two principal reasons. First, the analyst must distinguish between recently biosynthesized or native hydrocarbons and petroleum hydrocarbons in order to estimate petroleum contamination. The distinction can only be made by comparing hydrocarbons native to marine organisms with petroleum hydrocarbons to discern the characteristics of each group. Secondly, hydrocarbons in petroleum give such a wide range of molecular structures and molecular weights that no presently available method of analysis provides an accurate assessment of total petroleum contamination when considering all possible oil-contamination incidents.

*Petroleum Hydrocarbons* (see Figure 1). There are several classes of hydrocarbon in crude oils and most crude oil products. The distribution of hydrocarbons in these classes is briefly presented below. In general, the discussion of this subject by Blumer *et al.*[14] has been followed.

*n-Alkanes.* A series of n-alkanes extending from $C_1$ beyond $C_{60}$ is present, with a ratio of abundance of odd- to even-carbon-numbered chains of *ca.* 1.0. However, dewaxing removes most n-alkanes up to about n-$C_{40}$ from lubricating oils.

[14] M. Blumer, P. C. Blokker, E. G. Cowell, and D. F. Duckworth, in 'A Guide to Marine Pollution', ed. E. D. Goldberg, Gordon and Breach, New York, 1972, Ch. 2, 'Petroleum'.

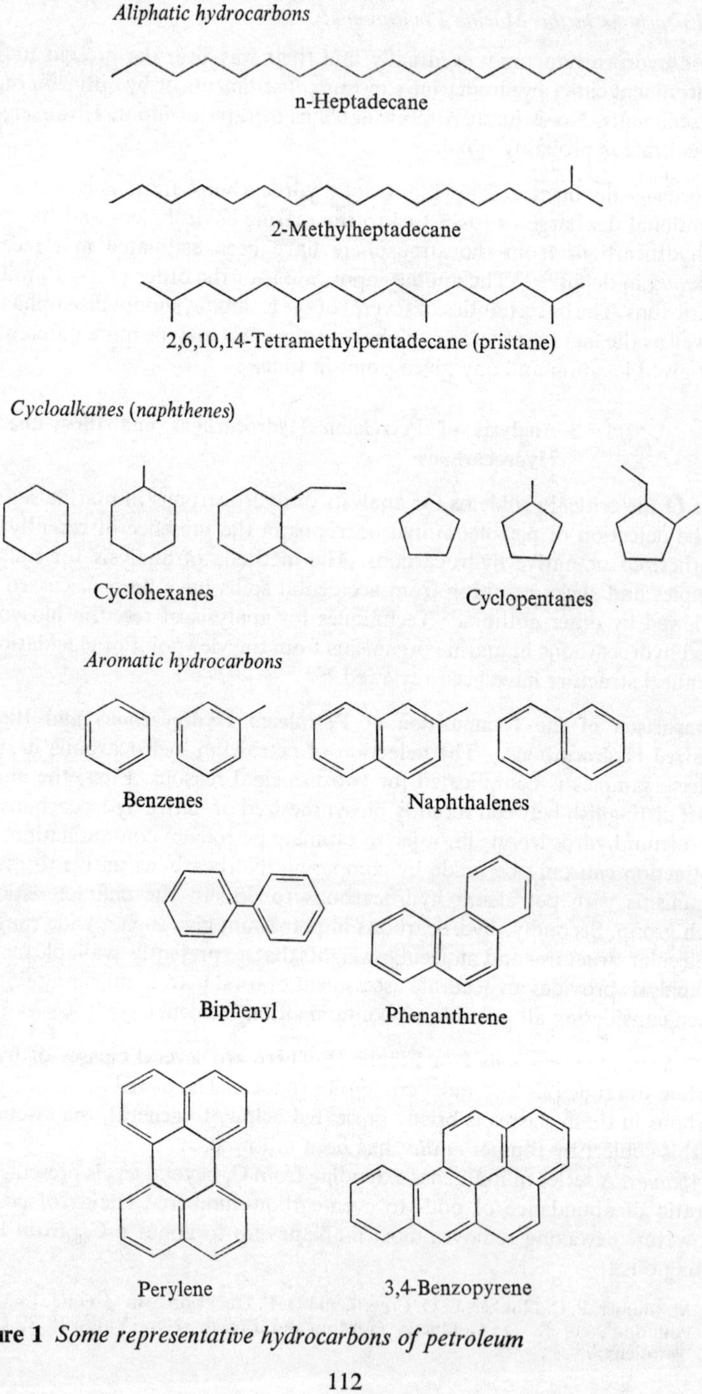


**Figure 1** *Some representative hydrocarbons of petroleum*

*Branched alkanes.* Many parallel homologous series of isomers, including parallel homologous series of isoprenoid alkanes, are present in petroleum.
*Cycloalkanes (naphthenes).* A complex mixture of compounds including substituted and unsubstituted rings is present, with the substituted compounds more abundant than their parent compounds.
*Aromatic hydrocarbons.* Again a complex mixture of compounds is present, including mono- and poly-alkylbenzenes, naphthalenes, and polynuclear aromatic hydrocarbons with multiple alkyl substitution. Also included in this class are compounds sometimes designated as naphtheno-aromatics owing to the mixed nature (aromatic, cycloalkane) of the subunits.
*Alkenes (olefins).* These compounds can be present in refinery products but are generally absent in crude oils.[14]

Several authors have discussed the hydrocarbon composition of crude oils and crude oil products in detail.[15]

*Recently Biosynthesized or Native Hydrocarbons* (see Figure 2). Some hydrocarbons in marine organisms may come from the ocean, either picked up from the water directly through permeable surfaces such as gills or obtained along with food through absorption from the gut.[16,17] Hydrocarbons are more stable than most biochemicals, so it would not be unexpected to find some derived from land sources in marine organisms. Organisms can also make their own hydrocarbons, either *de novo* or from precursor compounds

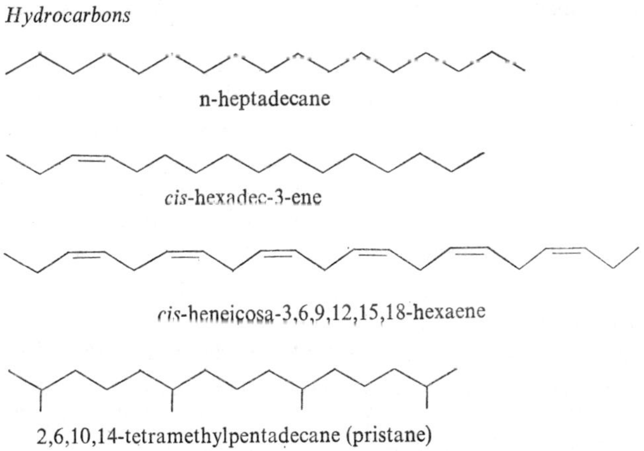


**Figure 2** *Some representative recently biosynthesized hydrocarbons*

[15] G. C. Speers and E. V. Whithead, 'Crude Petroleum, ref. 1. Ch. 27; M. A. Bestougeff, in 'Fundamentals of Petroleum Geochemistry', ed. B. Nagy and Colombo, Elsevier, 1967, Ch. 3, 'Petroleum Hydrocarbons'; H. M. Smith, 'Qualitative and Quantitative Aspects of Crude Oil Composition,' U.S. Dept. of the Interior, Bureau of Mines Bulletin, No. 642, 1968.
[16] R. F. Lee, R. Sauerheber, and A. A. Benson, *Science*, 1972, **177**, p. 344.
[17] M. Blumer, *Science*, 1967, **156**, 390.

obtained with their food. An example of the latter is the conversion of phytol into pristane.[18] Non-pollutant hydrocarbons from these sources are all of recent biosynthetic origin; they may all be called native to organisms.

The most recent comprehensive review of hydrocarbons native to organisms was that of Gerarde and Gerarde in 1961—62.[19] In 1966, Clark[20] reviewed the literature on saturated hydrocarbons (mostly non-cyclics).

*n-Alkanes*. Both land and marine organisms synthesize n-alkanes. Odd-carbon-number chains predominate, although even-carbon-number chains are present. In many instances one or two odd-carbon-chain n-alkanes predominate over all others. In marine phytoplankton, the n-alkanes with 15, 17, 19, and 21 carbon atoms are most abundant,[20] whereas in marsh grasses and *Sargassum* 21—29-carbon-atom n-alkanes predominate.[20,21] Some bacteria have been found to contain equal amounts of even-carbon-number and odd-carbon-number n-alkanes between n-$C_{25}$ and n-$C_{32}$.[22] Marine sponges and corals have also been shown to have little if any odd-carbon preferences for the $C_{25}$—$C_{34}$ n-alkanes.[23]

*Branched alkanes*. Several branched alkanes have been found in organisms. Pristane is the most abundant alkane in some fishes.[17] Phytane has not been detected in copepods from the Gulf of Maine nor in basking shark liver.[17] However, phytane has been reported to be present in bacteria.[24,25] A biochemical pathway for converting phytol into pristane but not into phytane has been demonstrated.[18] Several monomethyl branched alkanes[25,26] have been found in organisms and an alkylcyclopropane has been tentatively identified.[24,26]

*Alkenes* (*olefins*). Alkenes are often a major portion of the hydrocarbons found in organisms, particularly marine organisms. Squalene is the major hydrocarbon constituent of basking shark liver oil and cod liver oil. Isoprenoid $C_{19}$ and $C_{20}$ mono-, di-, and tri-olefins are present in copepods and some fishes.[27] Several straight-chain mono- to hexa-olefins have been found in marine organisms.[26—29] Carotenes are polyolefins and are found in considerable quantities in many organisms.

[18] J. Avigan and M. Blumer, *J. Lipid Res.*, 1968, **9**, 350.

[19] H. W. Gerarde and D. F. Gerarde, *Quart. Rev. Assoc. Food and Drug Officials U.S.*, 1961—62, **XXV**, **XXVI**.

[20] R. C. Clark, jun., M. S. Thesis, Department of Geology and Geophysics, Massachusetts Institute of Technology, Cambridge, Mass., 1966; M. Blumer, R. R. L. Guillard, and T. Chase, *Marine Biol.*, 1971, **8**, 183.

[21] K. Burns and J. M. Teal, *Deep Sea Res.*, 1973, **20**, 207.

[22] J. B. Davis, *Chem. Geol.*, 1968, **3**, 155.

[23] C. B. Koons, G. W. Jamieson, and L. S. Ciereszko, *Bull. Amer. Assoc. Petroleum Geologists*, 1965, **49**, 301.

[24] J. Han and M. Calvin, *Proc. Nat. Acad. Sci. U.S.A.*, 1969, **64**, 436.

[25] J. Sever Ph.D. Thesis, University of Texas, Austin, Texas, 1971.

[26] W. W. Youngblood, M. Blumer, R. R. L. Guillard, and F. Fiore, *Marine Biol.*, 1971 **8**, 190.

[27] M. Blumer, J. C. Robertson, J. E. Gordon, and J. Sass, *Biochemistry*, 1969, **8**, 4067.

[28] M. Blumer, M. M. Mullin, and R. R. L. Guillard, *Marine Biol.*, 1970, **6**, 226.

[29] W. W. Youngblood and M. Blumer, *Marine Biol.*, 1973, **21**, 163.

*Cycloalkanes and cycloalkenes.* Hydrocarbons containing one to three non-aromatic rings are present in several herbs and other terrestrial plants.[19] Most are classified as terpenoids because of their structural relationship to isoprene.

*Aromatic hydrocarbons.* There are several reports of aromatic hydrocarbons having been isolated from terrestrial plants and spices.[19] ZoBell[30] has recently reviewed reports regarding the occurrences of polynuclear aromatic hydrocarbons in the marine environment. Some of the work cited suggests that some polynuclear aromatic hydrocarbons may be synthesized by marine micro-organisms.[30] The aromatic hydrocarbons native to organisms contain at most one or two substituents.[19] Spectrophotometric and mass spectrometric analysis of hydrocarbons isolated from the marine mussel *Mytilus edulis* from one location did not detect any aromatic hydrocarbons.[16]

*Summary.* Analyses of marine and land organisms have shown that many different types of hydrocarbon are produced by organisms. However, an analysis of an individual species does not reveal more than a few members of each homologous series. Many times only one class of hydrocarbon and in a few cases only one individual hydrocarbon predominates. The only exceptions to this generalization might be the micro-organisms other than phyto-plankton, *i.e.* bacteria, yeasts, actinomycetes. Some of the reports of the analysis of hydrocarbons in bacteria[4,24,25] contain gas chromatograms which suggest that the saturated hydrocarbons have a wider range of molecular weight and structure than has been found in other organisms. There is a need for further analysis of these micro-organisms cultured under conditions which would preclude contamination by petroleum hydrocarbons.

It is important to note that only a limited number of terrestrial and marine species from only a few geographical locations have been analysed for their native hydrocarbons. The stage in the life cycle of the organism at the time of analysis, the culturing conditions, and the health of the organism should also be considered since such factors can influence the hydrocarbon distribution.[21,24,28]

Many investigators have limited their analytical techniques in order to search for only one or two classes of hydrocarbon—usually alkanes and alkenes—and would not have detected other classes, if present, or have not reported their presence. This is particularly true with respect to aromatic hydrocarbons and to a lesser extent cyclic alkanes.

*Characteristics of Petroleum Hydrocarbons Useful for Detecting Petroleum Contamination.* A comparison of the hydrocarbons in petroleum with hydrocarbons native to organisms shows the following differences which are useful in detecting petroleum contamination. Some of these differences may not apply for all organisms or all fossil fuels.

[30] C. ZoBell, *Proc. Joint Conference on Prevention and Control of Oil Spills*, 1971, American Petroleum Institute, Washington, D.C., p. 441.

(i) Petroleum contains a much more complex mixture of hydrocarbons with a much greater range of molecular structures and molecular weights than has been reported for hydrocarbons native to organisms.

(ii) Petroleum contains many homologous series. Adjacent members of a series usually are present in nearly the same concentration. The unity ratio of even- and odd-numbered n-alkanes is an example.

(iii) Petroleum contains numerous cycloalkanes and numerous aromatic hydrocarbons in comparison with the small number of each class which are native to marine organisms. In particular the numerous alkyl-substituted ring compounds have not been reported in organisms. Examples are the series of mono- and poly-alkylated benzenes and naphthalenes.

(iv) Petroleum contains numerous naphtheno-aromatic hydrocarbons. These compounds have not been found in organisms.

In addition to the differences in hydrocarbon composition noted above, petroleum contains hetero-compounds (S, N, O, or metals) and heavy asphaltic compounds. The numerous hetero-compounds have not been found in organisms, and heavy asphaltic compounds are not native to organisms.

There are situations where it is much more difficult to distinguish between petroleum hydrocarbons introduced by man and the hydrocarbons from natural sources. The analysis of water, organisms, or sediments taken from areas near natural oil seeps is one of the more difficult situations. It is possible that extensive chemical analysis would allow an estimate of the contributions from these two sources. In some cases it appears that it may be possible to identify sources for individual hydrocarbons. Recent analyses of the stereochemical isomers of pristane from two different sources suggest that this type of analysis may provide a clue as to the relative contributions from different sources.[3,31]

If *ca.* 0.5 g or more of hydrocarbons can be isolated from a sample, they can be dated using $^{14}C$ dating techniques. An age much older than Recent (10 000 years) indicates a predominance of petroleum hydrocarbons. This assumes that recently biosynthesized hydrocarbons did not have a fossil carbon source and that the contribution of hydrocarbons from weathering of ancient sediment is negligible. This technique has been used in the analysis of hydrocarbons isolated from surface sediments in a harbour area.[32]

**Sampling and Analysis.**—*Sample Contamination.* Samples taken in areas not immediately adjacent to shores must be obtained from a sampling platform, usually a ship. A ship is a floating localized source of hydrocarbons. Machinery is lubricated, bilges contain oily wastes which are periodically discharged, fumes are discharged from the exhaust funnel, and winch wires on which sampling gear is suspended are usually lubricated. Constant vigilance and special precautions are needed to obtain a sample not contaminated by the ship's environment. The same vigilance and precautions are needed to avoid

[31] R. E. Cox, J. R. Maxwell, R. G. Ackmann, and S. N. Hooper, *Canad. J. Biochem.*, 1972, **50**, 1238.

[32] O. C. Zafiriou, *Estuarine and Coastal Marine Science*, 1973, **I**, 81.

contamination if samples are analysed on board the ship. Controls involving all aspects of the extraction and analytical procedure should be processed to be certain that contamination has been avoided. Periodic samples of the ship's fuel, bilges, and lubricating oils should be obtained. Comparison of hydrocarbon composition of these potential contaminants with the hydrocarbons isolated from the samples will provide a means of determining if contamination has occurred.[33]

*Intercalibration and Comparison of Data.* Hydrocarbons in marine samples have been analysed by a number of laboratories employing several different extraction procedures. Comparison and integration of the results of these analyses to obtain a coherent view of hydrocarbon distribution in the marine environment assumes that the inter-laboratory accuracy and precision allows such a comparison and integration. A first attempt at the intercalibration of hydrocarbon analyses of marine samples has been reported.[34] Further and more extensive efforts are needed.

*Extraction.* Extraction of hydrocarbons from all marine samples has involved the same basic procedure. Organic solvents are used to extract hydrocarbons. The extracts also contain all or a portion of the lipid material present in the sample and require the further isolation of the hydrocarbons from the other lipid material.

Several extraction methods have been used to isolate hydrocarbons from organisms. Extraction with organic solvents using a Soxhlet apparatus, or using a homogenization procedure, has been described (*e.g.* refs. 34 and 35). Connel[36] extracted samples of fish with diethyl ether followed by steam distillation of the ether extract to obtain a volatile fraction, which in the particular case cited was primarily kerosene. Some analysts advocate digestion of the sample in alcoholic KOH followed by partitioning of the hydrocarbons and unsaponifiable lipids into a non-polar solvent such as hexane.[37,38] The advantages of the digestion procedure are said to be disruption of cells and more efficient extraction of the hydrocarbons from the sample. Steam distillation has also been used to extract hydrocarbons from fish.[39]

Particulate matter isolated from sea-water would probably be extracted using many of the procedures reported above. The isolation of the particulate matter from sea-water presents some problems. The filters employed should

33 J. W. Farrington, C. S. Giam, G. R. Harvey, P. L. Parker, and J. Teal (contributors) 'Marine Pollution Monitoring: Strategies for a National Program, ed. E. D. Goldberg, NOAA U.S. Dept. of Commerce, Washington, D.C., 1972.

34 J. W. Farrington, J. M. Teal, J. G. Quinn, T. Wade, and K. Burns, *Bull. Environ. Cont. Toxicol.*, 1973, **10**, 129.

35 P. R. Mackie, A. S. McGill, and R. Hardy, *Environ. Poll.*, 1972, **3**, 9.

36 D. W. Connel, *Marine Poll. Bull.*, 1971, **2**, 188.

37 J. W. Blaylock, P. W. O'Keefe, J. N. Roehar, and R. E. Wilding, *Proc. Joint Conference on Prevention and Control of Oil Spills*, 1973, American Petroleum Institute, Washington, D.C., 1973.

38 J. Greffard and J. Meury, *Cahiers Oceanographiques*, 1967, **XIX**, 457.

39 R. G. Ackmann and D. Noble, *J. Fisheries Res. Board Canada*, 1973, **30**, 711.

be thoroughly checked to be certain that they are not adsorbing hydrocarbons from the sea-water in addition to removing the particulate matter.

Solvents such as $CH_2Cl_2$, $CCl_4$, n-$C_6H_{13}$, n-$C_5H_{12}$, Freon 113, and stabilized $CHCl_3$ have been used to extract sea-water samples, including surface slicks.[2,34,40–49] It has been suggested that procedures employing an XAD-2 resin to isolate polychlorinated biphenyls from sea-water might also be used with some modifications to isolate hydrocarbons from sea-water.[50] Polyurethane foam plugs which have been used successfully for isolating chlorinated hydrocarbons from sea-water might also be used effectively for isolating hydrocarbons.[51]

Some of the procedures used for extraction of organisms are also useful for the extraction of sediment samples. Organic solvent extractions, usually aided by Soxhlet apparatus, sonication, or ball-milling, have been employed.[1,31,34,52–54] Elemental sulphur is present in most sediments and is co-extracted with the hydrocarbons. The sulphur is usually removed from the extract by reaction with copper.[33,55]

Extensive comparisons of the efficiency and selectivity of the various extraction procedures for marine samples have not been conducted and are needed.

*Saponification.* Most extracts, especially extracts of organisms obtained with the above methods, contain esters of fatty acids, *e.g.* waxes and glycerides. The esters often interfere with the isolation of alkenes and aromatic hydrocarbons. Saponification breaks the esters into fatty acid salts and alcohols which are easily removed, but transesterification of the existing esters to methyl or ethyl esters can sometimes occur.[56] This must be avoided to prevent interference. Complete saponification is confirmed by the absence of the

[40] E. M. Levy, *Water Res.*, 1971, **5**, 723; *ibid.*, 1972, **6**, 57.
[41] M. Barbier, D. Joly, A. Saliot, and D. Tourres, *Deep Sea Res.*, 1973, **20**, 305.
[42] A. Zsolnav, *Marine Chem.*, 1972—73, **1**, 127; *Deep Sea Res.*, 1973, **20**, 923.
[43] S. Carlberg and C. B. Skarstedt, *J. Cons. Int. Explor. Mer.*, 1972, **34**, 506.
[44] R. A. Hites and K. Biemann, *Science*, 1972, **178**, 158; B. R. Simonett, D. H. Smith, G. Eglinton, and A. L. Burlingame, *Arch. Environ. Contam. Toxicol.*, 1973, **1**, 193.
[45] E. Peake and G. W. Hodgson, *J. Amer. Oil Chemists' Soc.*, 1966, **43**, 215.
[46] D. C. Gordon, jun. and P. A. Michalik. *J. Fisheries Res. Board Canada*, 1971, **28**, 1912; P. D. Keizer and D. C. Gordon, jun., *ibid.*, 1973, **30**, 1039.
[47] M. Gruenfeld, *Environ. Sci. Technol.*, 1973, **7**, 636.
[48] R. A. Brown, T. D. Searl, J. J. Elliot, D. E. Brandon, and P. H. Monaghan, ref. 37, p. 505.
[49] T. M. Iliffe and J. A. Calder, *Deep Sea Res.*, 1974, in the press.
[50] G. R. Harvey, Tech. Report No. 72—86, Woods Hole Oceanographic Institution, Woods Hole, Mass., 02543, unpublished manuscript.
[51] H. D. Gesser, A. Chow, F. C. Davis, J. F. Uthe, and J. Reinke, *Analyt. Letters*, 1971, **4**, 883; J. F. Uthe, J. Reinke, and H. D. Gesser, *Environ. Letters*, 1972, **3**, 177.
[52] J. W. Farrington and J. G. Quinn, *Estuarine and Coastal Marine Sci.*, 1973, **1**, 71.
[53] F. S. Brown, M. J. Baedecker, A. Nissenbaum, and I. R. Kaplan, *Geochim. Cosmochim. Acta*, 1972, **36**, 1185.
[54] J. Sever and P. Haug, *Nature*, 1971, **234**, 447.
[55] M. Blumer, *Analyt. Chem.*, 1957, **29**, 1039.
[56] R. L. Glass, *Lipids*, 1971, **6**, 919.

absorption band for esters in infrared spectra of the material remaining after saponification.

*Separation of Hydrocarbons from other Lipids.* A variety of techniques has been used to separate hydrocarbons from the lipids coextracted by the procedures discussed above. Thin-layer chromatography and column chromatography, singly and in combination, have been routinely used effectively to isolate hydrocarbon fractions from extracts.[31–38,52–54,57–63] High-pressure liquid chromatography is a useful technique which should gain wider acceptance after further research and testing. Hites and Biemann[44] have successfully used it to separate aromatic hydrocarbons according to ring number. Karger *et al.*[64] have used 2,4,7-trinitrofluorenone-impregnated Corasil I high-pressure liquid chromatography columns to separate isomers of polynuclear aromatic hydrocarbons.

Gel permeation chromatography has been suggested as a technique which would be useful to the analysis of high molecular weight components of petroleum in marine samples.[14]

*Analysis of Hydrocarbons.* We shall discuss each technique as to its merits for detecting petroleum hydrocarbons in the presence of comparable or higher levels of native hydrocarbons.

*Infrared spectrometry.* The absorption frequencies of native hydrocarbons overlap or coincide with the absorption frequencies of petroleum hydrocarbons, with the possible exception of absorption frequencies for aromatic hydrocarbons of petroleum in the long-wavelength region. Thus, this technique shows little promise for detecting small quantities of petroleum hydrocarbons in the presence of hydrocarbons native to organisms.[65]

*U.v. absorption and u.v. fluorescence spectrometry.* These analytical techniques detect the presence of aromatic hydrocarbons. The use of u.v. absorption and u.v. fluorescence techniques as an indicator of oil contamination has been described.[40,46,66] They can be applied to lipid extracts and unsaponifiable lipid extracts without isolating the hydrocarbons from the lipids.[40,46,66] Appropriate attention must be directed towards the possible presence of native aromatic hydrocarbons and the quenching of the fluorescence emission.

[57] M. Ehrhardt, *Environ. Poll.*, 1972, **3**, 257.
[58] M. Blumer, G. Souza, and J. Sass, *Marine Biol.*, 1971, **5**, 195.
[59] M. Blumer, J. Sass, G. Souza, H. Sanders, F. Grassle, and G. Hampson, Tech. Report 70—44, Woods Hole Oceanographic Institution, Woods Hole, Mass., 1970, unpublished manuscript.
[60] M. Blumer and J. Sass, Tech. Report 72—19, Woods Hole Oceanographic Institution, Woods Hole, Mass., 1972, unpublished manuscript.
[61] K. Burns and J. Teal, Tech. Report 71—69, Woods Hole Oceanographic Institution, Woods Hole, Mass., 1971, unpublished manuscript.
[62] R. C. Clark, jun. and J. S. Finley, ref. 37, p. 161.
[63] H. J. Cahnmann and M. Kuratsune, *Analyt. Chem.*, 1957, **29**, 1312.
[64] B. L. Karger, M. Martin, J. Loheac, and G. Guiochon, *Analyt. Chem.*, 1973, **45**, 496.
[65] H. B. Mark, jun., T. C. Yu, J. S. Mattson, and R. L. Kolpack, *Environ. Sci. Technol.*, 1972, **6**, 833.
[66] V. Zitko and W. V. Carson, Tech. Report No. 217, Fisheries Research Board of Canada, Biological Station, St. Andrews, N.B., 1970.

The limitations of these techniques are as follows. There is little indication of the complexity or molecular weight range of the mixture. There is the possible overlap of the u.v. absorption spectra of highly conjugated native alkenes with the absorption spectra of the aromatic hydrocarbons. This limitation may be overcome by selectively hydrogenating the alkenes. Finally, no indication of the presence of alkanes or non-conjugated alkenes is given.

*Gas chromatography* (see Figure 3). This technique has been useful in analysing samples for petroleum contamination.[32,33–37,39,41,44,52–54,57–62,67,68] Hydrocarbons are tentatively identified by co-injection of authentic standards or by determining Kovats retention indices of peaks in the gas chromatograms.[29] The presence of an unresolved complex mixture of hydrocarbons in medium- and high-resolution gas chromatograms has been used as an indicator of the presence of petroleum hydrocarbons. This unresolved complex mixture of hydrocarbons with a wide molecular weight range has not been detected in organisms free of petroleum hydrocarbons, with the possible exception of bacteria as previously discussed.

In several cases further analysis by g.c.–m.s. was employed to confirm identifications.

*Mass spectrometry and gas chromatography–mass spectrometry*. Hydrocarbons isolated from marine samples contaminated by petroleum have been analysed by g.c.–m.s. and by mass spectrometry alone.[32,41,44,48,57,60,67,68] This is particularly effective for detecting petroleum contamination when low-voltage mass spectrometry is used to analyse the aromatic fractions isolated from a sample. The mass spectra obtained show nominal mass numbers of fragments of aromatic hydrocarbons, *e.g.* in the case of aromatic hydrocarbons from a sample contaminated with petroleum, fragments of the series of alkylated aromatic hydrocarbons of a distribution and complexity similar to that found in petroleum itself are present in the mass spectra.

*Computer-interfaced mass spectrometry and gas chromatography–mass spectrometry*. The data-manipulating ability of a computer interfaced with a mass spectrometer or a g.c.–m.s. system creates a potentially powerful analysis system for identifying petroleum contaminants in the marine environment. The mass spectral data might be analysed by compound-type analysis,[69,70] although validation of these techniques for the analysis of petroleum hydrocarbons in the presence of equal or greater quantities of native hydrocarbons has not been documented. Recent applications of these techniques for the analysis of hydrocarbons isolated by column chromatography from marine sample extracts have been described.[41,44,48,67]

[67] M. Tissier and J. L. Oudin, ref. 37, p. 205; M. Tissier and J. L. Oudin, *Adv. Org. Geochem.*, 1973, in the press.

[68] R. C. Lao, R. S. Thomas, and J. L. Monkman, *Internat. J. Environ. Analyt. Chem.*, 1972, **1**, 187.

[69] A. Hood and M. J. O'Neil, in 'Advances in Mass Spectrometry', ed. J. D. Waldron, Pergamon Press, New York, 1959.

[70] D. H. Smith, R. W. Olson, F. C. Walls, and A. L. Burlingame, *Analyt. Chem.*, 1971, **43**, 1796; D. H. Smith and G. Eglinton, *Nature*, 1972, **235**, 325; D. H. Smith, *Analyt. Chem.*, 1972, **44**, 536.

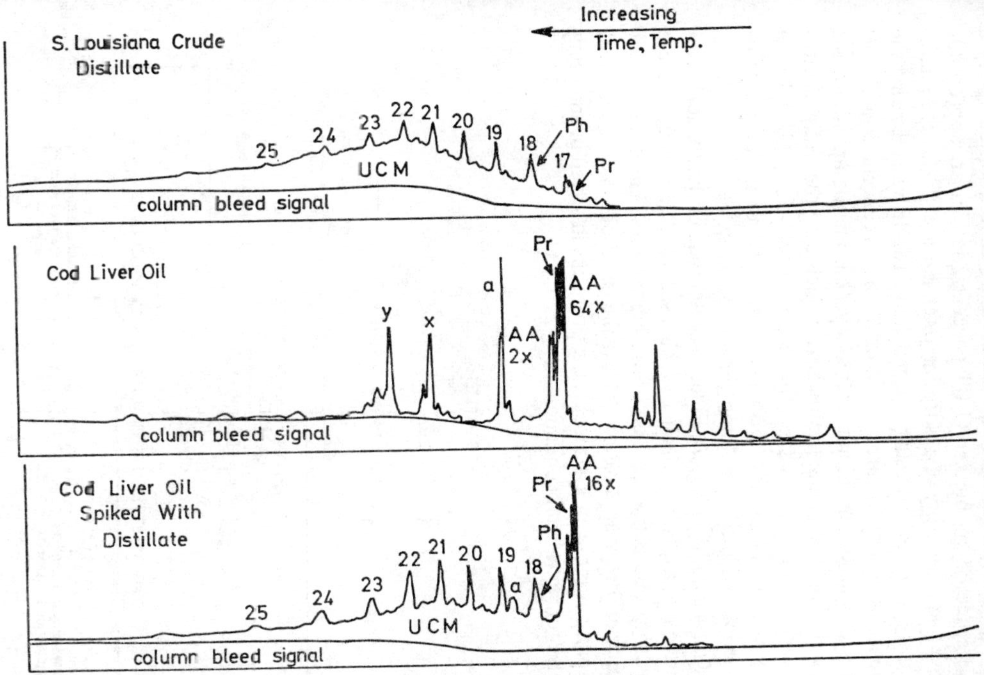


**Figure 3** *Gas-chromatographic analysis of hydrocarbons to detect petroleum.* AA: *automatic attenuation of detector signal;* Pr: *pristane;* Ph: *phytane; numbers: n-alkanes;* a: *phytadienes;* x, y: *unknown hydrocarbons;* UCM: *unresolved complex mixture. Conditions:* 3% *Apiezon L,* 2.3 m *on chromasorb W* 80/100 *mesh; carrier gas* $N_2$ *at* 10—15 ml $min^{-1}$; *the column was temperature programmed from* 80 *to* 290 °C *at* 6 °C $min^{-1}$

*Quantification.* The measurement of the concentrations of hydrocarbons in the various types of sample has been accomplished by a variety of techniques. Spectrometric quantitative analyses are made using calibration curves of a representative crude oil or bunker C oil concentration in an appropriate solvent plotted *versus* intensity of absorption or emission.[40,46] Gas-chromatographic analyses are quantified using internal standards or by calibrating the detector response using a hydrocarbon mixture of known concentration.[33,34,37] Quantities of hydrocarbons in column or thin-layer chromatography fractions have been determined using analytical balances, including the Cahn electrobalance for very small quantities of the order of 10—500 $\mu$g.[14,32–34,58,61] Spectrometric measurements have also been employed to quantify hydrocarbons in fractions isolated by column chromatography.[48]

*Application of the methods of analysis.* Figure 4 presents an outline of analytical techniques presently available which can be used to detect and estimate petroleum hydrocarbon contamination of marine samples. The development of new methods and new applications of existing methods can be expected to proceed rapidly as more analysts tackle the problems involved in analysing marine samples for petroleum contaminants.

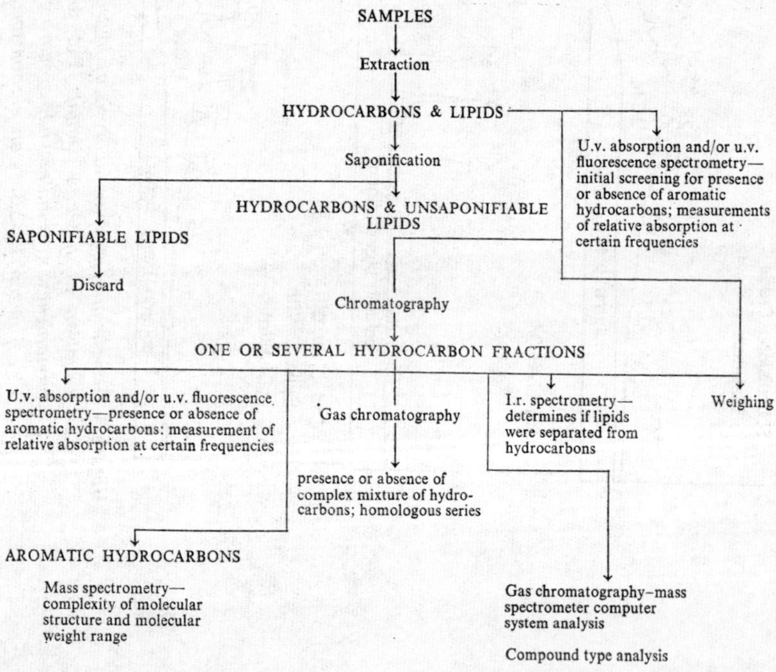


**Figure 4** *Flow diagram for analytical techniques to detect and estimate petroleum contamination*

These techniques provide information in each of the categories of characteristics for petroleum hyrocarbons described earlier. Gas chromatography provides information about the complexity and the molecular weight range. It also detects the presence or absence of certain homologous series, *e.g.* 1:1 ratio of even- and odd-carbon n-alkanes. In addition, gas chromatography provides some information about native hydrocarbons and their absolute and relative abundances.

The n-alkanes from petroleum may be hard to detect in the presence of native n-alkanes such as are found in bacteria, marsh plants, sponges, and corals. The absence of n-alkanes may mean that the sample does not contain petroleum contamination. This can be confirmed by the absence of other characteristics for petroleum hydrocarbons, or it could mean that the petroleum contaminants have been bacterially altered prior to or after incorporation into the organism. n-Alkanes are much more susceptible to bacterial attack than are other petroleum hydrocarbons (see later section).

The techniques also provide information about the presence or absence of aromatic hydrocarbons and their structural complexity and molecular weight range. Thus, we believe that careful application of these techniques will provide information about the presence or absence of petroleum contamination in marine samples and also provide an estimate of the severity of contamination.

*Reporting results of analyses.* The following information should be included when reporting results of analyses of marine organisms for petroleum contamination: (i) precision and accuracy of the method employed; (ii) limits of detection at the precision and accuracy stated; (iii) molecular weight range and molecular structure range over which the methods are precise and accurate; and (iv) molecular weight range and molecular structure range of the reported petroleum contamination in the sample.

## 4 Distribution of Hydrocarbons

**Marine Organisms.**—We have reviewed most of the literature concerned with hydrocarbons found in marine organisms in an earlier section.[1,2,4,14,16–34] Little research has been focused towards the understanding of the biological role of hydrocarbons in marine organisms. However, the role of squalene as an intermediate in the biosynthesis of steroids is described in the biochemical literature, as is the role of the carotene pigments. It has also been suggested that squalene, which is present in high percentages in the livers of deep-sea sharks, assists the sharks in achieving buoyancy.[71] This has also been suggested as a biological role for pristane in certain species of zooplankton.[72] The higher molecular weight $C_{25}$, $C_{27}$, $C_{29}$, and $C_{31}$ carbon n-alkanes found in the marsh grasses probably act as a waxy outer protective layer.

[71] R. M. Love, 'The Chemical Biology of Fishes', Academic Press, London, 1970, p. 143.
[72] M. Blumer, M. M. Mullin, and D. W. Thomas, *Helgolaender Wiss. Meeresuntersuch.*, 1964, **10**, 187.

The work of Blumer and co-workers and others indicates that hydrocarbons can be used as a chemotaxonomic tool and as a tool to elucidate food chains.[17,18,20–29,72–76] In one study[73] evidence was presented that hydrocarbons can be used as markers to distinguish subpopulations of fish which feed in different locations and therefore may ingest different assemblages of hydrocarbons. Several of the same workers have pointed to the restraints placed on the chemotaxonomic value of hydrocarbons by the variation of hydrocarbon composition with the growth stage of organisms.[20,27]

The pathways for biosynthesis of hydrocarbons in marine organisms deserve further attention. It has been established that pristane in certain zooplankton species (calanoid copepods) is derived from the phytol moiety of chlorophyll of the ingested phytoplankton.[18] The correlation of the structure of the $C_{22}$ fatty acid docosahexaenic acid and the $C_{21}$ hydrocarbon heneicosahexaene in certain phytoplankton species suggests an inter-relationship between fatty acid biosynthesis and hydrocarbon biosynthesis.[28] However, heneicosahexaene is present in benthic algae, and attempts to find the $C_{22}$ fatty acid in three species of brown benthic algae have failed to detect the fatty acid.[29] This suggests that no close inter-relationship between fatty acid biosynthesis and hydrocarbon biosynthesis exists in some cases.

It has been observed that the percentage of males produced in laboratory-reared *Calanus helgolandicus*, a marine zooplankton, depends on the algal composition of their food.[77] Youngblood *et al.*[24] commented on the intriguing correlation between the percentage of males produced and the predominance of heneicosahexaene in the algal food. These workers later demonstrated[29] that the alkene:alkane ratio is linked to the stage of tissue development in the benthic alga *Ascophyllum nodosum*.[29] Their comment that . . . 'we should no longer consider all plant hydrocarbons as accidental by-products of the cell biochemistry' offers a challenge to further research.

**Sea-water.**—A wide variety of analytical techniques has been utilized for analyses of sea-water.[2,40–46,48,49] This makes it difficult to compare results. A large proportion of the data available has been obtained using u.v.-fluorescent analysis of lipid extracts of sea-water.[40,46] Infrared spectrometry has been used to quantitate hydrocarbons from an extract using column chromatography.[43] Other studies utilized a microdetector for changes in heats of adsorption in the effluent from a liquid chromatograph column to measure amounts of hydrocarbons, and all quantities were determined relative to hydrocarbons in a reference sea-water sample.[42]

At best, all of the aforementioned studies provide a measure of some

[73] M. Blumer, M. M. Mullin, and D. W. Thomas, *Science*, 1973, **140**, 974.
[74] M. Blumer and D. W. Thomas, *Science*, 1965, **147**, 1148.
[75] K. Winters, P. L. Parker, and C. Van Baalen, *Science*, 1969, **158**, 467.
[76] R. F. Lee, J. C. Nevenzel, G. A. Paffenhöfer, A. A. Benson, S. Patton, and T. E. Kavanagh, *Biochim. Biophys. Acta*, 1970, **202**, 386.
[77] G. A. Paffenhöfer, *Helgolaender Wiss. Meeresuntersuch.*, 1970, **20**, 346.

portion of the various types of hydrocarbon present in sea-water, and extrapolate to the total concentration of hydrocarbons. The limitations of the i.r. and u.v.-fluorescence measurements have been previously discussed. Three studies have used column chromatography to isolate hydrocarbons from lipids extracted from sea-water and analysed one or more of the isolated hydrocarbon fractions by gas chromatography.[2,41,45,48,49]

Two of the studies analysed some of the hydrocarbon fractions by mass spectrometry or g.c.–m.s.[41,48] The mixture of hydrocarbons found in these cases was complex and had a wide molecular weight range. The results for several samples have been interpreted as showing that some of the hydrocarbons are from petroleum sources[48] or are consistent with but do not necessarily indicate petroleum sources.[41,48,49]

One study analysed a relatively large number of samples from the Western North Atlantic and the Gulf of Mexico.[48] There was a considerable geographical variability in the concentrations and composition of hydrocarbons. This variability also was apparent in another study in the Eastern Gulf of Mexico.[49] The reported concentrations of hydrocarbons have varied within the range 1.0—100 $\mu g\,l^{-1}$ for all of the above studies combined. Squalene and another polyunsaturated hydrocarbon, heneicosahexaene, have been isolated from surface sea-water in a coastal region.[2] Other studies of hydrocarbons in sea-water have not reported the presence of olefins or have presented no definitive data to demonstrate their presence or absence. In several of the studies sea-water samples were filtered before analysis of the water and for the rest of the studies they were not. A comparison of the data is hampered by this variable. There are no data available on the concentrations of hydrocarbons on particulate matter and in filtered sea-water taken from the same sample.

In most studies there was little attention given to analysing control samples to ensure that water samples were not contaminated during the sampling and analysis operations, a problem which is particularly hard to avoid for sea-water samples collected and fully or partially extracted at sea. It does not suffice to run controls on the analytical procedures and solvents used in a shore-based laboratory if some of the extractions were carried out at sea.

Based on the variability of hydrocarbons among species and subpopulations of marine organisms, and on the biochemical stability of hydrocarbons, Blumer and co-workers suggested that when released to the water column hydrocarbons would serve as unique signatures for the various water masses of the ocean.[2,72] Thus far this hypothesis has not been rigorously tested. The hypothesis has been restated recently.[41,49] It is interesting to note that with the exception of coastal water samples[2] the analyses of water samples previously cited have not allowed unambiguous correlations of hydrocarbon composition with the hydrocarbon composition of phytoplankton and zooplankton living in the water mass. This may be a function of petroleum pollution or hydrocarbons from other micro-organisms such as bacteria and yeast obscuring the hydrocarbons contributed by the plankton, or it may be the result of petroleum hydrocarbon contamination during sampling and analysis.

**'Tarballs' and Tar Particles.**—We offer only a brief synopsis of a recent extensive discussion of this subject[78] and comment on some immediate research needs. The present data on the distributions of tar in the ocean are mainly collected from the Western North Atlantic Ocean. There is a need for collecting more extensive data on tar distribution in the world's oceans. Of particular importance is the extension of surveys for tar particles to the waters underlying the surface layers of the ocean. This would establish whether or not tar particles sink from surface waters into the deep waters and are eventually deposited as sediments. Recently, small bits of tar have been observed in a few cores of sediment taken in 1500 m of water in the Gulf of Mexico.[79]

Analyses of the chemical composition of tar are for the most part limited to gas chromatography analysis, fractionation by column chromatography, or metal analysis. Thus, there are data available about the molecular weight range or boiling-point distribution of tar constituents and the relative proportions of classes of compounds and metals. The distribution of n-alkanes and branched alkanes has been determined for a large number of tar particles. Detailed analyses of the composition of the cyclic alkanes, aromatic hydrocarbons, and heteroatom compounds are not yet available, but should be forthcoming in the near future as a result of ongoing research.

Chemical analysis of tar particles strongly indicates that the major source of the tar in the Western North Atlantic is from tanker ballast discharges.[78] Analyses of tar particles in the Western Gulf of Mexico suggest that some of the tar particles in this location have an oil-seep source.[80]

**Slicks**—Analyses of hydrocarbons in slicks not associated with known oil-pollution incidents have been reported.[81,82] A comparison of the concentration of hydrocarbons in the surface microlayer and in water 20 cm below the surface taken in coastal waters showed some enrichment in the surface microlayer.[81] Identification of individual hydrocarbons was hampered by low concentrations of 5.9—8.5 $\mu g\, l^{-1}$ and small sample size. Several surface

78 J. N. Butler, B. F. Morris, and J. Sass, 'Pelagic Tar from Bermuda and the Sargasso Sea', Bermuda Biological Station for Research, Special Publication No. 10, 1973.

79 L. M. Jeffrey, D. J. Frank, N. Powell, A. Bautz, A. Vos, and L. May, 'Progress Report on Pelagic, Beach, and Bottom Tars of the Gulf of Mexico and Controlled Weathering Experiments', Dec. 1973, Texas A & M University, Department of Oceanography, College Station, Texas.

80 C. B. Koons, in 'Background Papers for a Workshop on Inputs, Fates, and Effects of Petroleum in the Marine Environment', U.S. National Academy of Sciences, Washington, D.C., Vol. II, 1973, p. 495.

81 R. A. Duce, J. G. Quinn, C. E. Olney, S. R. Piotrowicz, B. J. Ray, and T. L. Wade, *Science*, 1972, **176**, 161.

82 J. G. Quinn and T. L. Wade, 'Background Papers for the Workshop on Baseline Studies of Pollutants in the Marine Environment', Brookhaven National Laboratory, May, 1972, Office for the International Decade of Ocean Exploration, National Science Foundation, Washington, D.C., p. 633.

microlayer and 20 cm subsurface samples from the North Atlantic have been analysed.[82] The composition of the hydrocarbons was variable and within the molecular weight range of the 16—30-carbon n-alkanes. Concentration values ranged from 5 to 55 $\mu g\, l^{-1}$ for surface microlayer samples and from 10 to 56 $\mu g\, l^{-1}$ for subsurface samples with no consistency of enrichment of hydrocarbons in the surface microlayer.

**Surface Sediments.**—There have been several studies devoted to the analysis of hydrocarbons in ancient sediments and sediments less than 10 000 years old which were buried and not in contact with the contemporary marine environment. The reports in the literature have been reviewed.[1,3,4]

Analyses of surface sediments in some coastal areas free from petroleum pollution or natural oil seeps have shown that $C_{25}$, $C_{27}$, $C_{29}$, and $C_{31}$ n-alkanes predominate in the higher molecular weight fraction of hydrocarbons. These n-alkanes presumably have their origin in land plants and marsh-grass waxes. Branched alkanes, including the isoprenoid hydrocarbon pristane, have been reported for surface sediments.[83] Sever[25] reported some very interesting hydrocarbon distributions for surface sediments in coastal and lagoonal areas of the Gulf of Mexico. She found a wide-ranging distribution of 21—30-carbon n-alkanes with a predominance of odd-carbon n-alkanes, the branched iso- and anteiso-17- and -18-carbon alkanes, and the 7- and 8-methylheptadecanes. She concluded that the composition of the hydrocarbons in these surface sediments pointed towards blue-green algae and higher plants as a primary source for the hydrocarbons.

Surface sediments of a marine lagoon on Surtsey, a recently formed volcanic island, contained a hydrocarbon mixture suggesting a blue-green algal origin.[54] There was an even distribution of the 17—30-carbon n-alkanes. The isoprenoid hydrocarbons pristane and phytane were reported. Pristane and phytane were also reported in surface sediments from a reducing environment in a coastal region[53] and from surface sediments of the Dead Sea.[84] Whereas pristane has been reported to be present in surface sediments from several locations, phytane has not been found in some surface sediments.[53,83]

There are a few reports of the presence of alkenes in surface sediments. In some cases gas-chromatographic analyses of hydrocarbons isolated from sediments have shown that one or two alkenes of undetermined structure are the predominant hydrocarbons in the surface sediments.[52,85,86] Heneicosahexaene and heptadecene have been reported to be present in surface sediments.[86]

The diagenesis of hydrocarbons in a salt marsh environment has been studied by Johnson and Calder.[87] They concluded that *in situ* microbial

[83] M. Blumer and W. D. Snyder, *Science*, 1965, **150**, 1588.
[84] A. Nissenbaum, M. J. Baedecker, and I. R. Kaplan, *Geochim. Cosmochim. Acta*, 1973, **36**, 709.
[85] J. W. Farrington, Ph.D. Thesis, University of Rhode Island, 1972.
[86] M. Ehrhardt and M. Blumer, *Environ. Poll.*, 1972, **3**, 179.
[87] R. W. Johnson and J. A. Calder, *Geochim. Cosmochim. Acta*, 1973, **37**, 1943.

activity caused major changes in the composition of hydrocarbons, and that these changes occur rapidly and are of large magnitude. The changes appeared to be similar but not exactly the same for both reducing and oxidizing depositional environments. Changes in hydrocarbon composition and concentration in sediment samples from a reducing environment in another geographical location, and further offshore, were not as marked.[53]

The cycloalkanes have not received the attention accorded the n-alkanes and branched alkanes. Their presence in sediments containing polluting petroleum hydrocarbons has been noted by the unresolved complex mixture in the gas chromatograms of the alkanes isolated from the sediments.[32,52,59,88,89]

Aromatic hydrocarbons have also not been subjected to the same extensive study that has been accorded the n-alkanes and branched alkanes. The exceptions to this generalization are 3,4-benzopyrene and perylene. There has been a review of the analyses for 3,4-benzopyrene in marine samples which includes several papers on analyses of surface sediments.[90] The dominant method of analysis has been, until recently, separation of a hydrocarbon fraction containing the 3,4-benzopyrene and then analysis of this fraction by u.v.-fluorescence spectrometry. There are some problems with the discriminatory power of some of these methods of analysis and we think that application of more sophisticated techniques, such as further separation of the 3,4-benzopyrene from other interfering compounds and analysis by mass spectrometry and u.v.-fluorescence spectrometry, will be needed to provide more accurate concentration values for surface sediment samples.

The reports of perylene in surface sediments have recently been reviewed by Aizenshtat.[91] Several mechanisms have been postulated for the chemical formation of perylene from 4,9-dihydroxyperylene-3,10-quinone and from erythroaphin pigments in reducing environments. This is a restatement of an earlier theory based on more extensive analysis. Two recent papers report analyses of aromatic hydrocarbons in surface sediments of the French Coast.[67] Concentrations of 3,4-benzopyrene,8,9-benzofluoranthene, perylene, and fluoranthene varied from not detectable to 13.0 $\mu$g (100 g)$^{-1}$ dry weight for each hydrocarbon. In addition, there appeared to be a series of alkylated aromatic hydrocarbons. The more complex mixtures of aromatic hydrocarbons occurred at the places where there was usually a higher concentration. This was interpreted as indicating petroleum pollution in the surface sediments, an interpretation supported by the extreme complexity and wide molecular weight range of the saturated hydrocarbons.[67]

Several papers have dealt with the detection of petroleum pollution in surface sediments.[32,33,52,59–61,67,68,88,89] In general, the extreme complexity and wide molecular weight range of hydrocarbons, along with high concentrations

[88] M. Blumer and J. Sass, *Marine Poll. Bull.*, 1972, **3**, 92.
[89] M. Blumer and J. Sass, *Science*, 1972, **176**, 1120.
[90] J. B. Andelman and M. J. Suess, *Bull. World Health Organization*, 1970, **43**, 479.
[91] Z. Aizenshtat, *Geochim. Cosmochim. Acta*, 1973, **37**, 559.

of hydrocarbons, have been interpreted to demonstrate petroleum pollution in the sediments. The rationale supporting this interpretation is discussed along with other possibilities which might lead to the complex mixture of hydrocarbons in sediments with a wide molecular weight range.[14,32,34,52,60,88,89] A detailed study of hydrocarbons in sediments polluted by a No. 2 fuel oil spill—West Falmouth, Mass. Oil Spill—has been conducted over a two-year period and is to date the most extensive and comprehensive study of the detection of petroleum hydrocarbons in surface sediments and the fate of petroleum hydrocarbons once incorporated into surface sediments.[60,88,89] Some of the conclusions are as follows. The fuel oil penetrated to a depth not exceeding 7.5 cm. Partial dissolution and bacterial degradation altered the composition of the oil near the sediment–water interface. However, oil was still present in the sediment for two years after the spill and demonstrated that oil products and crude oils can have a considerable environmental persistence.

The persistence of a portion of petroleum hydrocarbons from a spill was also documented for oil coming ashore on a beach at Martha's Vineyard, Mass. and oil coming ashore on Bermuda.[92]

**Marine Atmosphere.**—A few samples of the marine atmosphere taken from a tower on Bermuda have been analysed for higher molecular weight hydrocarbons. Normal alkanes, branched alkanes, and a mixture of hydrocarbons not resolved by medium-resolution gas chromatography, probably cyclic alkanes and branched cyclic alkanes, were reported for the molecular weight range between that of the $C_{14}$ and $C_{32}$ n-alkanes. A review of the scanty information available concerning hydrocarbons in the marine atmosphere was also presented in that paper.[93]

**Concentrations of Hydrocarbons in Sea-water, Organisms, and Sediments.**—The usual ranges of concentrations of hydrocarbons in the marine environment are given below. We present these in a separate section to facilitate a comparison of the concentration levels in the various types of sample. The values are taken from references cited in the previous sections.

*Sea-water.* Concentrations of 1—100 $\mu g\, l^{-1}$ or $10^{-9}$ parts per part of total hydrocarbons are usually found. Individual hydrocarbons may be present at the ng $l^{-1}$ level. Higher concentrations have been reported in areas near industrial effluents and municipal sewage effluents. In general, surface waters contain higher concentrations than deeper waters of the ocean.

Tar particle concentrations in the upper 1 m of the water column range from 0 to 40 mg $m^{-2}$ with up to 500 mg $m^{-2}$ reported for some areas of the Mediterranean Sea. Thus, tar particle concentrations in surface waters are of the order of 0—40 $\mu g\, l^{-1}$, which represents a substantial portion of the total hydrocarbon concentration in some surface water samples.

[92] M. Blumer, M. Ehrhardt, and J. H. Jones, *Deep Sea Res.*, 1973, **20**, 239.
[93] R. A. Duce, J. G. Quinn, and T. L. Wade, *Marine Poll. Bull.*, 1974, **5**, 59.

*Organisms*. Hydrocarbon concentrations in organisms range from 1 to 200 $\mu$g $g^{-1}$ wet weight or $10^{-6}$ parts per part for total organisms. Higher concentrations are present in certain organs; *e.g.* shark liver samples have concentrations up to 1300 $\mu$g $g^{-1}$ wet weight. In some cases one or two hydrocarbons account for almost all the hydrocarbon content of a sample.

*Sediments*. Hydrocarbon concentrations of 1—100 $\mu$g $g^{-1}$ dry weight or $10^{-6}$ parts per part are found for sediments in the continental margin areas, *i.e.* continental slope and continental shelf. The sediments in estuarine areas may have higher concentrations but usually range from 5 to 100 $\mu$g $g^{-1}$ dry weight. There are no values available in the literature for surface sediments below the 3500 m water depth.

*Oil-polluted Samples*. Hydrocarbon concentrations in oil-polluted samples range up to 545 $\mu$g $g^{-1}$ wet weight for organisms in coastal waters, and up to 12 400 $\mu$g $g^{-1}$ dry weight for sediments in coastal areas. Concentrations in water samples associated with spills are high, but there is no way of separating the total concentration into a dissolved portion and an emulsified portion. Values of up to 1.0 mg $l^{-1}$ have been reported for oil in water not immediately adjacent to a slick.

**Processes Controlling the Distribution of Hydrocarbons.**—*Physical-chemical.* The mechanisms of input into the marine environment—biosynthesis, effluent discharges, accidental oil spills, atmospheric fallout, *etc.*—have an obvious influence on the distribution of hydrocarbons. The mechanism of input also determines whether the hydrocarbons enter at the fresh water–salt water interface, the air–sea interface, or the sediment–water interface or are directly introduced as an integral component from organisms. The input processes also determine whether the hydrocarbons enter in a dispersed form or in a cohesive manner such as an oil slick.

Quinn and co-workers[94–96] have recently reported a series of studies on the adsorption–desorption of alkanes, aromatic hydrocarbons, and No. 2 fuel oil on minerals and the solubilization of hydrocarbons in sea-water. Hydrocarbons can be removed from marine waters by adsorption on to mineral particles and incorporation into the underlying sediments. This association process can occur by two routes: contact of the hydrocarbons and mineral particles in the upper parts of the water column achieved during the settling of the mineral down past the floating or suspended hydrocarbons, or transport of the hydrocarbon by water circulation to the bottom. There the hydrophobic hydrocarbon can adhere to the bulk sediments. Regardless of the process involved it is unlikely that 100% of the hydrocarbons become associated with the sediments. Many of the hydrocarbons have small but measurable solubilities in water, so that at least a fraction will dissolve in sea-water rather

[94] P. A. Meyers, Ph.D. Thesis, University of Rhode Island, 1972.
[95] P. Boehm and J. G. Quinn, *Geochim. Cosmochim. Acta*, 1973, **37**, 2459.
[96] P. A. Meyers and J. G. Quinn, *Nature*, 1973, **244**, 23.

than adsorb on to solid surfaces. The differences in relative solubilities of the various types of hydrocarbon—normal alkanes, branched alkanes, aromatic—can lead to the distribution in sediments observed in laboratory and field studies of petroleum-polluted sediments.[32,52,60,67,88,89] A study of a No. 2 fuel oil spill[60] showed that after two years aromatic hydrocarbons exhibited preferential losses of lower molecular weight components from marine sediments while more highly substituted aromatics, which would be more insoluble, were preferentially retained. Uptake and retention by marine sediments of n-alkanes and branched alkanes was greater than that of relatively more soluble cyclic and aromatic components of petroleum in a study of adsorption of petroleum from water.[94,96]

Several factors can influence the solubility of hydrocarbons in sea-water and thus modify their incorporation into sediments. Water temperature and salinity would affect uptake by sediment. The presence of certain types of dissolved organic compound in sea-water can enhance the solubility of hydrocarbons.[45,94] Fulvic acid is an example. Poorly soluble hydrocarbons can associate with these relatively soluble materials and become distributed in sea-water in true solution or in finely dispersed micelles. This tends to diminish their adsorption by sediment particles.[94–96] Laboratory studies have shown that removal of indigenous organic matter results in a significant increase in the ability of sediments to adsorb hydrocarbons.[94,96] However, because of a high surface area per unit of sediment rising from a small average grain size or clay mineral content, the sediments which have adsorbed more natural organic matter also tend to adsorb more hydrocarbons than other sediments. Thus, a correlation between total organic content and hydrocarbon uptake is observed.

*Biological.* Hydrocarbons biosynthesized by organisms are released to sea-water or sediments by excretion from living organisms or by autolysis and decomposition of dead organisms.

The uptake, retention, and discharge of hydrocarbons by living organisms has received considerable attention in the recent literature because of its importance to the question of biological effects of oil pollution and the potential public health hazard posed by oil-polluted marine food which might be consumed by man. The earlier work of Blumer and co-workers suggested that hydrocarbons, particularly pristane, were passed up the food web from prey to predator.[14,72] Analysis of oil-polluted shellfish before and after placing the shellfish in clean sea-water in a laboratory showed very little loss of the petroleum hydrocarbons after at least 60 days.[59]

Four recent studies[16,97–99] have demonstrated that there is an active uptake of hydrocarbons from sea-water across the membrane system of shellfish and fish. Presumably the primary site of uptake is the gill membranes based

[97] J. J. Stegeman and J. M. Teal, *Marine Biol.*, 1973, **22**, 37.
[98] J. W. Anderson, ref. 80, p. 690.
[99] Publication No. 4191, ed. B. E. Vaughan, American Petroleum Institute, Washington, D.C., 1973.

on studies using $^{14}C$-labelled hydrocarbons.[16] Within marine organisms there are varying concentrations of hydrocarbons in tissues and organs. Lipid-rich organs such as the liver and reproductive organs have higher concentrations of hydrocarbons than relatively lipid-poor tissues such as muscles.[17]

The composition of petroleum hydrocarbons in animals exposed to No. 2 fuel oil was definitely different than the hydrocarbon composition of the No. 2 fuel oil in several studies.[59,97–99] Aromatic hydrocarbons were taken up to a greater extent than were saturated hydrocarbons.[97–99] One study showed the opposite,[16] a result which may be linked with the fact that the organisms were not simultaneously exposed to both saturated hydrocarbons and aromatic hydrocarbons.

The studies[16,97–99] also showed that petroleum hydrocarbons taken up by shellfish were discharged to a significant extent when placed in clean water for periods of up to 52 days. However, in some cases there were still detectable levels of petroleum hydrocarbons after several weeks of exposure to clean water.[16,97] It is obvious that several processes and factors are operative in the relationship between hydrocarbons in organisms and hydrocarbons in their surrounding habitats. The findings of Blumer and co-workers are not necessarily in contradiction to the findings of the other workers. The former studied shellfish which had been exposed over a two month period to a massive dose of oil in their habitat from an oil spill whereas the latter exposed shellfish to soluble or 'accommodated' oil in sea-water in laboratory experiments. Stegeman and Teal[97] have summarized all of these studies in their statement, 'We suggest that the concentration and composition of non-biogenic hydrocarbons in oysters reflects both the complete exposure history and the current exposure level, and results from the interplay between uptake, release, retention, route of entry, and possible residence of hydrocarbons in multiple (lipid?) compartments'.

Clearly, this is an area of active research with the promise of some significant advances in our understanding of the relationship between organic compounds in sea-water, their uptake by, retention by, effect on, and release by marine organisms.

## 5 Fate of Hydrocarbons in the Marine Environment

There are several mechanisms for the removal of hydrocarbons from the contemporary marine environment. These are photochemical and/or chemical alteration to other compounds, biochemical alteration to other compounds, transfer to the atmosphere, or burial in sediments as geological deposits. The relative importance of the processes depends on the hydrocarbon or mixture of hydrocarbons in question. The lower molecular weight hydrocarbons ($C_1$—$C_{10}$) would be influenced to a greater extent by evaporation and solubilization processes. Photo-oxidation would be expected to have a greater effect on aromatic hydrocarbons and heteroatom aromatics than on n-alkanes and branched alkanes, although the photochemistry of complex mixtures of organics is a difficult process to predict. Although auto-oxidation is certainly to be expected in spilled oil, its relative rate may be much slower than the other processes acting on the oil.

**Incorporation into Sediments.**—Incorporation into sediments by mixing processes, deposition of hydrocarbons adsorbed on to particulate matter or incorporated into particulate matter, is one fate of hydrocarbons in the marine environment. The processes have been discussed earlier. Biochemical degradation of the hydrocarbons after incorporation into sediments may be much slower than in the water column or at the air–sea interface because of lower oxygen concentrations or anaerobic conditions in the sediments.[60,88,89] Hydrocarbons surviving the degradation process in the surface sediments are eventually buried and become geochemical deposits. Studies of the similarities in chemical structure between hydrocarbons found in ancient sediments and hydrocarbons synthesized by organisms offer strong evidence that hydrocarbons survive for millions of years once incorporated into deeply buried sediments.[1,3,4]

**Transfer to the Atmosphere.**—Hydrocarbons from oil slicks and hydrocarbons from biosynthetic processes incorporated into natural sea slicks may enter the atmosphere by evaporation or by injection during turbulence, or *via* bubbles bursting through the surface film[81,82,100,101] There has been little published on this particular fate. Once in the atmosphere the relative amounts that are chemically or photochemically degraded and the amount which is redeposited on to the oceans or land are unknown.

**Biochemical Alteration.**—The microbial degradation of oil pollutants and specifically hydrocarbons in the marine environment has been recently reviewed.[102] Earlier reviews of petroleum microbiology and microbial degradation of oil in the marine environment are also recommended.[103,104] In general, studies have shown that n-alkanes are more easily degraded than are the other types of hydrocarbon. Branched alkanes are the next in order of ease of biodegradation. Cyclic alkanes and aromatic hydrocarbons seem to be much more resistant to biochemical degradation by micro-organisms.

A two year study of petroleum hydrocarbons incorporated into sediments from an oil spill has demonstrated that the above order of biochemical degradation is the same in a natural marine system.[60,89]

It has been established that the ability of micro-organisms to degrade hydrocarbons depends on several factors including temperature, pH, nutrients, presence of other organic compounds, previous exposure of the

[100] H. G. Widmark, W. D. Garret, and K. H. Palmork, ref. 14, Ch. 4.
[101] D. C. Blanchard, *Advances in Chemistry*, ed. R. Baier, 1974, in the press.
[102] 'The Microbial Degradation of Oil Pollutants', ed. D. G. Ahearn and S. P. Meyers, Publication No. LSU-SG-73-01, 1973, Center for Wetland Resources, Louisiana State University, Baton Rouge, Louisiana.
[103] J. B. Davis, 'Petroleum Microbiology', 1967, Elsevier, New York.
[104] C. E. ZoBell, *Proc. Joint Conference on Prevention and Control of Oil Spills*, 1969, American Petroleum Institute, Washington, D.C., p. 317.

culture to various types of hydrocarbon, and physical agitation or mixing. Detailed studies of the microbial degradation of hydrocarbons, particularly cyclic alkanes and aromatic hydrocarbons, in natural systems are needed to further our understanding of the processes and rates of degradation of hydrocarbons in sea-water and sediments.

Bacteria capable of partially degrading petroleum have been isolated from several locations in the world's oceans.[102,105] However, the rate at which hydrocarbons are degraded in the open ocean has not yet been ascertained.

The products of biochemical degradation of hydrocarbons have been recently summarized. The principal products resulting from the microbial degradation of hydrocarbons are various hydroperoxides, alcohols, phenols, carbonyls, aldehydes, ketones, and esters in addition to $CO_2$ and $H_2O$.[102] The exact pathways of biochemical degradation are poorly understood.

There are a few reports of studies of the metabolism of hydrocarbons by marine animals—both fish and invertebrates—using radioisotope-labelled hydrocarbons.[16,106] Animals were exposed to sea-water containing the labelled hydrocarbons. Thus, the studies did not deal with metabolism of hydrocarbons introduced *via* feeding but rather those taken up across body membrane surfaces such as the gills. It was demonstrated that certain fish can metabolize the aromatic hydrocarbons naphthalene and 3,4-benzopyrene, with the major metabolites tentatively identified as 7,8-dihydro-7,8-dihydroxybenzopyrene and 1,2-dihydro-1,2-dihydroxynaphthalene. However, the marine mussel *Mytilus edulis* did not metabolize n-heptadecane, tetrahydronaphthalene, naphthalene, or 3,4-benzopyrene when tested under similar conditions.[106] The studies have been thus far limited in the length of time of observation and the species investigated. Nevertheless, they offer a model for more extensive investigations.

## 6 Oil Spills

The fates of actual and simulated oil spills have been studied with varying degrees of scientific sophistication. The exact fate of each spill is unique since it is influenced by several environmental parameters such as temperature, salinity, wind and wave conditions, suspended sediment load, and geographical location and by the chemical composition of the spilled oil.[13,92,107] The studies as a whole point to the following processes as important influences on the fate of oil: dissolution, evaporation, photo-oxidation, adsorption on to particulate matter, mixing below the surface, which often includes mixing into sediments in shallow areas, adsorption on to rocky shores and beaches, and microbial attack and degradation.[13,92]

[105] O. G. Mironov, *Atti. 5° Coll. Int. Oceanogr. Med.*, Messina, 1973, p. 315.
[106] R. F. Lee, R. Sauerheber, and G. H. Dobbs, *Marine Biol.*, 1972, **17**, 201.
[107] D. Straughan, *J. Petrol. Technol.*, 1972, 250.

Gruenfeld[108] and Zafiriou *et al.*,[109] among others, have recently reviewed the principles of passive tagging or fingerprinting of petroleum contamination—the correlation of a particular case of petroleum contamination with its source or sources of contamination by primarily chemical parameters.

There is an extensive literature on the subject of oil spills, oil pollution, and its effects and fate. Several compendiums of references are available.[110–112] For those readers who wish a detailed listing of all literature dealing with oil pollution, references 113 and 114 are suggested. A recent study sponsored by the U.S. National Academy of Sciences concerning the 'Input, Fate and Effect of Oil in the Marine Environment' should be issued by the time this Report is published.[114]

## 7 Summary

The study of hydrocarbons in the marine environment is now progressing at a faster pace owing to the interest in understanding the inputs, fate, and effects of oil pollution. However, it is clear that a more complete knowledge of the hydrocarbons introduced into the marine environment by natural processes such as biosynthesis is needed.

All studies of hydrocarbons are dependent on the analytical techniques employed, and our knowledge of hydrocarbons in the marine environment was, until recently, hampered by a lack of a more widespread application of sophisticated methods of analysis. Because of their ease of analysis and their predominance in most marine samples, n-alkanes, n-alkenes, and branched alkanes have been studied to greater extent than other types of hydrocarbons. A more detailed knowledge of the cyclic and branched cyclic alkanes and naphthenoaromatic and aromatic hydrocarbons in marine samples is essential. Many of the important questions relating to oil-pollution studies and to chemical processes in surface sediments will be answered by such knowledge.

There have been significant advances in the past few years in understanding the interactions between hydrocarbons and minerals, and dissolved organic matter and hydrocarbons. These studies, when expanded and coupled with similar studies of other organic compounds in the marine environment, will

108 M. Gruenfeld, ref. 37, p. 179.
109 O. C. Zafiriou, M. Blumer, and J. Meyers, Tech. Report No. 72—55, Woods Hole Oceanographic Institution, Woods Hole, Mass., 02543, 1972, unpublished manuscript.
110 A. Nelson-Smith, Oil Pollution and Marine Ecology, Paul Elek, Scientific Books, Ltd., London, 1972.
111 D. S. Moulder and A. Varley, 'A Bibliography on Marine and Estuarine Oil Pollution, The Laboratory of the Marine Biological Association of the United Kingdom, Plymouth, Devonshire, U.K., 1971.
112 *Proceedings of Joint Conference on Prevention and Control of Oil Spills*, American Petroleum Institute, Washington, D.C.; Conferences held in 1969, 1971, and 1973.
113 'Oil Pollution: An Index Catalog to the Collection of the Oil Spill Information Center', University Library, University of California, Santa Barbara, California.
114 'Inputs, Fates and Effects of Oil in the Marine Environment', A Report of a Workshop Convened by the Ocean Affairs Board, U.S. National Academy of Sciences, Washington, D.C., 1975, (in the press).

provide us with a more detailed knowledge of the processes governing the distribution of organic compounds.

The chemical and biological processes acting on hydrocarbons in surface sediments are being explored. Concurrently, the role of sediment micro-organisms as a source for hydrocarbons in surface sediments is being mentioned with increasing frequency. Only a few studies have demonstrated that micro-organisms contribute substantially to the hydrocarbons in surface sediments. We expect that more evidence will be forthcoming in the near future. We also expect that during the next few years it will become more apparent that chemical and biological processes are operating in surface sediments and have a significant influence on which types of organic compound are found in ancient sediments.

There is a dynamic and complicated interaction between hydrocarbons in organisms and hydrocarbons in the surrounding environment. Furthermore, key studies have established a relationship between concentration and composition of hydrocarbons in organisms and the growth stage or stage of tissue development. The door to understanding the biochemical role of hydrocarbons has begun to open, with great promise for significant advances in this area of research. Concurrently, the pathways of biosynthesis of hydrocarbons will be unravelled.

There has been a significant advance in our understanding of the physical, chemical, and biological processes affecting oil spilled into the marine environment, particularly with respect to the biodegradation of oil by micro-organisms. This knowledge is far from complete, and we cannot predict with certainty the fate of any given oil spill. However, it is certain that petroleum hydrocarbons can and do linger long after visual evidence of the oil spill has disappeared.

We thank Dr. J. G. Quinn and Dr. M. Blumer for valuable comments on sections of the manuscript. We also thank Drs. J. M. Hunt, O. C. Zafiriou, and J. Stegeman for their comments. The authors wish to acknowledge financial support received during the period of preparation of this Report, which includes:

to J. W. Farrington, from the Environmental Protection Agency, Grant 802724; from the Office for the International Decade of Ocean Exploration, National Science Foundation, Grant GX 35212; from the Office of Naval Research, Contract NOOO14-66-CO241; NRO83-004;

to P. A. Meyers, from the donors of the Petroleum Research Fund, administered by the American Chemical Society, partial support.

# 6
# The Fate of DDT and PCB's in the Marine Environment

BY M. M. RHEAD

## 1 Introduction

*p,p'*-DDT or 2,2-bis-(*p*-chlorophenyl)-1,1,1-trichloroethane (1) was first synthesized by Zeidler in 1874, but it was not until 1939 that Muller[1] discovered its insecticidal properties. A comprehensive history of this compound has been written by Metcalf.[2]

Pure *p,p'*-DDT melts at 108—109 °C, has a vapour pressure at 20 °C of $1.5 \times 10^{-7}$ mmHg, and has a solubility in water[3] of 1.2 parts in $10^9$. It is considerably more soluble in organic solvents.[4] In the presence of strong alkali, *p,p'*-DDT is dehydrochlorinated to the non-insecticidal *p,p'*-DDE or 2,2-bis-(*p*-chlorophenyl)-1,1-dichloroethylene (2). DDE melts at 85 °C and has a solubility in water almost identical to that of DDT[2] (*i.e.* 1.3 parts in $10^9$).

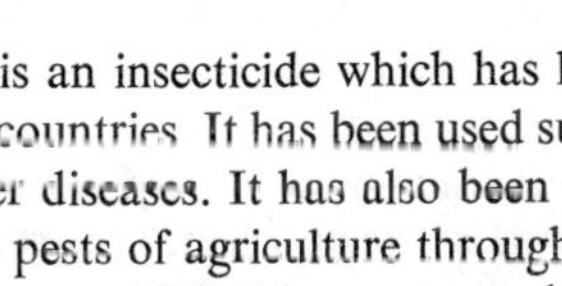


DDT (1)

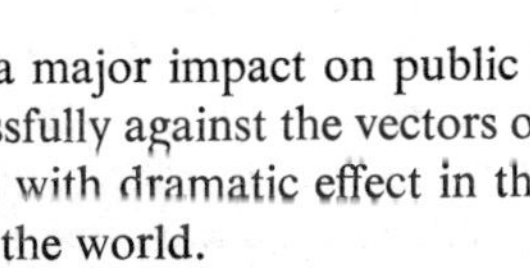


DDE (2)

DDT is an insecticide which has had a major impact on public health in tropical countries. It has been used successfully against the vectors of malaria and other diseases. It has also been used with dramatic effect in the control of insect pests of agriculture throughout the world.

Major use of DDT began around 1944. Between then and 1968 a total of $1.2 \times 10^{12}$ g had been manufactured in the United States alone.[5] Between

[1] 'Das Insectizid Dichlorodiphenyl Trichloroethane und Seine Bedeutung', ed. P. Muller, Vol. I, Birkhauser-Verlag, Basel, Switzerland, 1955.
[2] R. L. Metcalf, *J. Agric. Food Chem.*, 1973, **21**, 511.
[3] M. C. Bowman, J. Acree, and M. K. Corbett, *J. Agric. Food. Chem.*, 1960, **8**, 406.
[4] R. L. Metcalf, in 'Pesticides in the Environment', Part I, ed. R. White-Stevens, Marcel Dekker, New York, 1971, pp. 1—144.
[5] E. D. Goldberg (Chairman), 'Chlorinated Hydrocarbons in the Marine Environment', A report prepared by the Panel on Monitoring Persistent Pesticides in the Marine Environment, National Academy of Sciences, Washington D.C., 1971.

1961 and 1966 annual U.S. production averaged $6 \times 10^{10}$ g. Risebrough[6] has speculated that world annual production of DDT over this period averaged about $10^{11}$ g.

The class of compounds denoted by the abbreviation PCB was first described in 1881 by Schmidt and Shultz, and first recognised as commercially useful in 1930 by Penning.[7] PCB's are prepared by chlorinating biphenyl. They are not obtained as single compounds but as mixtures of compounds of the same basic structure with different numbers of chlorine atoms in different positions (3).

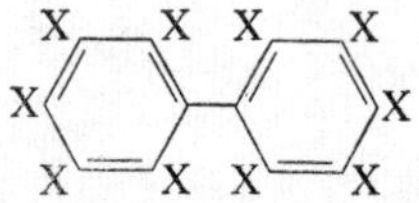


PCB (3) (X = Cl or H)

Although theoretically there could be 210 different isomers in a PCB mixture, in practice fewer isomers are encountered. Mass spectrographic studies of Arochlor 1260 (a PCB mixture produced commercially by the Monsanto Company and containing 60% chlorine by weight) shows the presence of only 11 isomers; 5 containing six chlorines, 5 containing seven chlorines, and 1 containing eight chlorine atoms.[8] In a similar study[9] Arochlor 1254 was found to consist of 18 isomers.

PCB's are apparently almost insoluble in water but very soluble in organic solvents, although figures are not available. They are chemically inert, being unaffected by strong alkalis, acids, or corrosive chemicals.[10] PCB's have low volatilities (though not as low as DDT); thus, a mixture containing mainly pentachlorobiphenyls had a vapour pressure at 30 °C of $6 \times 10^{-5}$ mmHg, while a similar tetrachlorobiphenyl mixture had a vapour pressure (30 °C) of $3 \times 10^{-4}$ mmHg.[11]

Estimated uses of PCB's in various fields are as follows:[12] 45—50% of total production in electrical condensers, 15% as insulating material in electrical transformers, 10—15% for heating media, 10—15% as solvents for non-carbon copying paper, 5—10% for plasticizers, and the remainder for a multitude of minor uses. More details of usage may be found elsewhere.[10,12] Large amounts of PCB are often recovered as waste when, for example, an electrical appliance is scrapped. The discarded items are often disposed of by

[6] R. W. Risebrough, in 'Impingement of Man on the Oceans', ed. D. W. Hood, Wiley-Interscience, New York, 1971, pp. 259—286.
[7] C. H. Penning, *Ind. and Eng. Chem.*, 1930, **22**, 1180.
[8] J. H. Koeman, M. C. Ten Noever de Brauw, and R. H. Vos, *Nature*, 1969, **221**, 1126.
[9] G. E. Bagley, W. L. Reichel, and E. Cromarter, *J. Assoc. Offic. Analyt. Chemists*, 1970, **53**, 251.
[10] D. B. Peakall and J. L. Lincer, *Bioscience*, 1970, **20**, 958.
[11] S. Tanaka, *J. Jap. Tech. Assoc. Pulp and Paper Ind.*, 1971, **27**, 3 (Fisheries Research Board of Canada, Translations Series No. 2527).
[12] R. Tatsukawa and N. Isono, *Shizen* (*Nature*), 1971, **6**, 30 (Fisheries Research Board of Canada, Translations Series No. 2259).

burning, throwing into water, or by burial. All of these methods of disposal may contribute to environmental pollution by PCB's. PCB's do not fully decompose on burning but may enter the atmosphere in significant quantities. Some manufacturers are reported[12] to dispose of PCB's by heating above 1000 °C but even this treatment may result in incomplete conversion into carbon dioxide and hydrogen chloride.

Among the companies manufacturing PCB's are Monsanto in the United States and Britain (trade name Arochlor), Prodelèc, France (Phenochlor), Bayer, Germany (Colphen), Kanegafuchi, Japan (Kanechlore), and Mitsubishi-Monsanto, Japan (Arochlor and Santotherm), and they are also made in the Soviet Union (Soval). Little information is available of the amounts of PCB produced by these companies. It is said[12] that Kanegafuchi Chemical Company produces 1000 tons per month and Mitsubishi-Monsanto 400 tons per month.

## 2 Laboratory Studies of Biological Degradation of DDT and PCB's

The degradation of pesticides by a wide range of organisms, including birds, fish, and micro-organisms, has been reviewed by Menzie.[13] More-specific reviews will be referred to in the relevant parts of this section.

**Aquatic Plants.**—Cox[14] has recently reviewed the uptake and metabolism of DDT in marine phytoplankton.

Small conversions of DDT into DDE in marine phytoplankton species have been reported.[14—17] Keil and Prester[15] estimated a 5 to 10% conversion of DDT to DDE in *Cylindrotheca closterium* cultures after 21 days. Rice and Sikka[16] studied uptake and metabolism of DDT in the species *Skeletonema costatum*, *Cyclotella nana*, *Isochrysis galbana*, *Olisthodiscus luteus*, *Amphidinium carteri*, and *Tetraselmis chuii* of marine algae. They observed conversion of DDT into DDE in all species. Conversion ranged from 0.03 to 11% after 24 days treatment. One species, *Skeletonema costatum*, produced, in addition to DDE, a small amount of an unknown polar metabolite from DDT. Bowes[17] studied the metabolism of DDT in seven species of marine phytoplankton, representative of five algal divisions. He recovered small amounts, *ca.* 3.0—7.4%, of DDE from cultures of *Skelatonema costatum*, *Cyclotella nana*, *Thalassiosira fluviatilis*, and *Dunaliella tertiolecta* after 2 to 3 weeks. He noted that, in the case of *D. tertiolecta*, DDE was only recovered after rupture of the cells. He suggested that resistance of this species to DDT

[13] C. M. Menzie, Special Scientific Report, Fish and Wildlife Services, No. 127, U.S. Department of Interior, Washington, D.C., 1969.
[14] J. L. Cox, *Residue Rev.*, 1972, **44**, 23.
[15] J. E. Keil and L. E. Prester, *Bull. Environ. Contam. Toxicol.*, 1969, **4**, 169.
[16] C. P. Rice and H. C. Sikka, *J. Agric. Food Chem.*, 1973, **21**, 148.
[17] G. W. Bowes, *Plant Physiol.*, 1972, **49**, 172.

toxicity involved penetration of the chloroplast and subsequent conversion into DDE. Laboratory cultures of fresh-water diatoms (a *Nitzschia* species and an unidentified species) have been reported[18] to metabolize DDT, forming DDE only, but the amount of conversion was slight.

**Fish.**—Hatchery-reared Atlantic salmon (*Salmo salar*) degraded an unspecified proportion of DDT, absorbed from aqueous suspensions, to DDE and 2,2-bis-(*p*-chlorophenyl)-1,1-dichloroethane, DDD (4), within 9 hours. DDT was also adsorbed onto the outer surfaces of the salmon but was not degraded at these sites.[19] *In vitro* studies[20] using fresh mid- and hind-gut contents of *Salmo salar* and DDT showed conversion into DDD but no DDE. The conversion was attributed to intestinal microflora. Wedemeyer[21] suggested that most of the metabolism of DDT in fish resulted from the action of intestinal microflora which converted DDT into DDE and/or DDD. Studies with rainbow trout, *Salmo gairdneri*, showed that the liver converted DDT into DDE.[21] Bullhead fish have converted DDT into DDMU or 2,2-bis-(*p*-chlorophenyl)-1-chloroethylene (5).[22] *In vitro* studies with carp

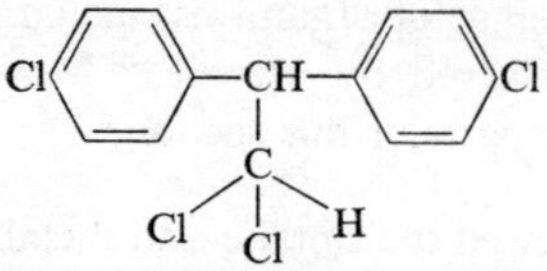

DDD (4)

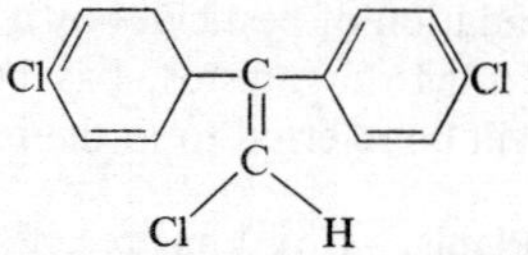

DDMU (5)

blood and DDT indicated some conversion into DDD, DDE, and possibly DDMU.[23] More recently, Ernst reported rapid uptake of DDT by the flatfish *Platichthus flesus*[24] and *Solea solea*,[25] which both gave DDD (10% conversion by *S. solea*) and smaller amounts of DDE but no polar metabolites.

**Micro-organisms.**—The degradation of DDT and related compounds in systems dominated by micro-organisms has been extensively studied. A variety of microbes have been used under anaerobic and aerobic conditions as pure cultures or as mixed cultures from natural sources. The subject has

[18] S. Miyazaki and A. J. Thorsteinson, *Bull. Environ. Contam. Toxicol.*, 1972, **8**, 81.
[19] G. L. Greer and U. Paim, *J. Fisheries Res. Board Canada*, 1968, **25**, 2321.
[20] A. D. Charrington, U. Paim, and O. T. Page, *J. Fisheries Res. Board Canada*, 1969, **26**, 47.
[21] G. Wedemeyer, *Life Science*, 1968, **7**, 219.
[22] R. Bridges, B. J. Kallman and A. K. Andrews, *Trans. Amer. Fisheries Soc.*, 1963, **92**, 421.
[23] D. D. Cruca, Bureau of Sport, Fisheries, and Wildlife Resources Publication No. 64, 1968, p. 129.
[24] W. Ernst, *Veröffentl. Inst. Meeresforsch. Bremerhaven*, 1970, **12**, 353.
[25] W. Ernst, *Veröffentl. Inst. Meeresforsch. Bremerhaven*, 1970, **12**, 361.

recently been reviewed by a number of authors.[26–29] Investigations of the degradation of DDT by micro-organisms are listed in Table 1.

The product most widely reported to result from DDT degradation by micro-organisms is DDD (4). The compounds DDE (2), DDMU (5), DDMS (6), DDNU (7), DDA (8), DPM (9), DBH (10), DBP (11), kelthane (12), BA (13), and others have been reported as minor products. A major new compound, DDCN (14), has been reported[30,31] as a main product (*ca.* 10% yield[30]) of the degradation of DDT in anaerobic sewage sludge.

**Table 1** *Microbial degradation of DDT (The conditions used are described in the text)*

| *Authors* | *Reference* | *Organism* | *Products* |
|---|---|---|---|
| Wedemeyer | 35 | *Aerobacter aerogenes* | DDD, DDE, DDMU, DDMS, DDNU, DDA, DPM, DBH, DBP |
| Plimmer *et al.* | 45 | *Aerobacter aerogenes* | DDD |
| Mendel *et al.* | 36 | *Aerobacter aerogenes* | DDD |
| Mendel and Walton | 41 | *A. aerogenes, E. coli* | DDD |
| Barker *et al.* | 42 | *Proteus vulgaris* | DDD, DDMU, DDMS |
| Fries *et al.* | 49 | Rumen micro-organisms | DDD |
| Chacko *et al.* | 38 | Soil *Actinomycetes* | DDD |
| Patil *et al.* | 39 | Soil micro-organisms | DDD, Kelthane, DDA |
| Guenzi and Beard | 32 | Soil micro-organisms | DDD, DDE, Kelthane, DDA, DBP, DPM, BA |
| Albone *et al.* | 30 | Sewage sludge | DDD, DDCN |
| Jensen *et al.* | 31 | Sewage, Lake sediment | DDCN |
| Engst and Kuyawa | 48 | *Fusarium oxysporum* | DDD, DDMU, DDOH, DDA, DBP |
| Patil *et al.* | 51 | Marine micro-organisms | DDD, DDE, DDOH, DDNS, Unknown polar compounds |
| Pfaender and Alexander | 54 | *Hydrogenomonas* sp. | DDD, DDMS, DDMU, DDE, DBH, DPM, DDA, PCPA |
| Pfaender and Alexander | 56 | Sewage micro-organisms | DDD, DDE, DBP |

26 G. F. Fries, Advances in Chemistry Series, No. 111, 1972, p. 256.
27 A. W. Bourquin, 'The Microbial Degradation of Oil Pollutants,' ed. D. G. Ahearn and S. P. Meyers, Louisiana State University, Center for Wetland Resources, Louisiana, 1973, 237—243.
28 G. T. Brooks, in 'Fate of Pesticides in the Environment,' ed. A. S. Tahori, International IUPAC Congress on Pesticide Chemistry, Tel Aviv (1971), 1972, p. 223.
29 R. M. Pfister and F. Matsumura, in 'Critical Reviews in Microbiology', ed. A. I. Laskin and H. Lechevalier, The Chemical Rubber Co., Cleveland, Ohio, vol. 2, 1972, pp. 1—33.
30 E. S. Albone, G. Eglinton, N. C. Evans, and M. M. Rhead, *Nature*, 1972, **240**, 420.
31 S. Jensen, R. Göthe, and M. O. Kindstedt, *Nature*, 1972, **240**, 422.

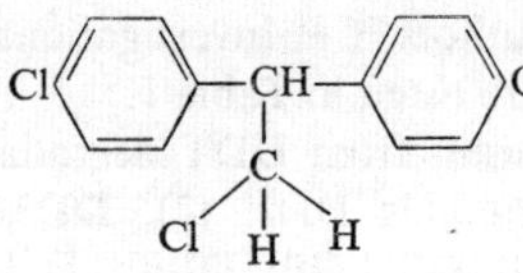

DDMS (6)

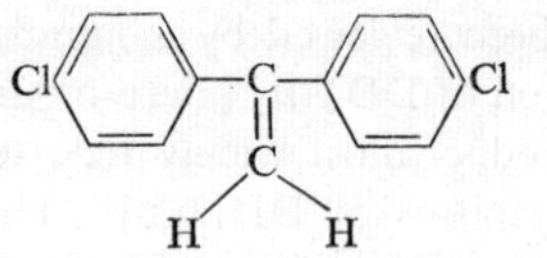

DDNU (7)

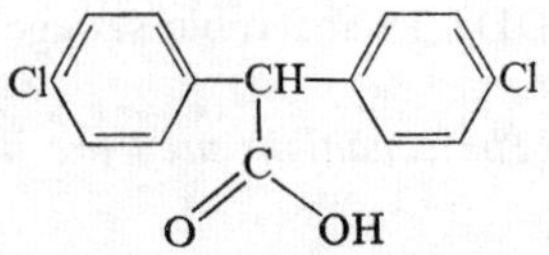

DDA (8)

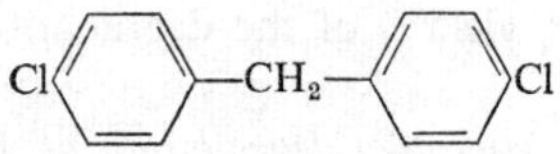

DPM (9)

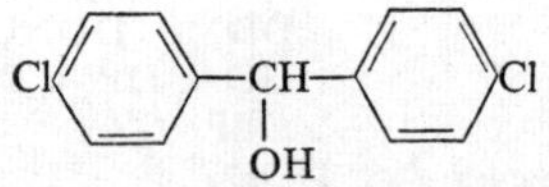

DBH (10)

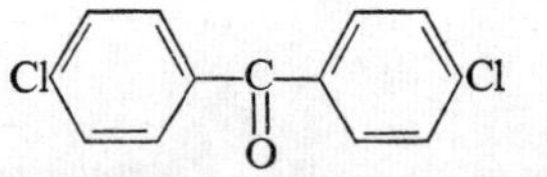

DBP (11)

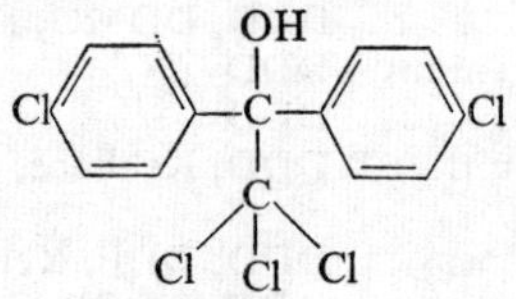

Kelthane
(12)

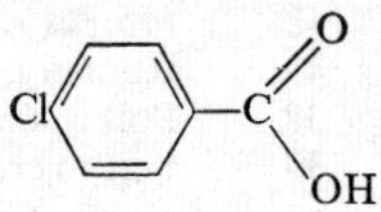

BA (13)

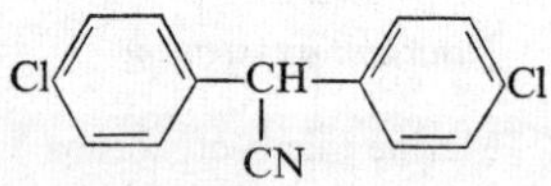

DDCN (14)

The authors in one report[30] speculated that DDCN might be formed by a biological mechanism or a chemical mechanism. If DDCN is produced chemically, as seems the more likely possibility, then reducing conditions are probably essential. The role of the organisms in sewage sludge might then be to ensure the maintenance of the appropriate conditions and thereby indirectly effect the conversion of DDT into DDCN.

There is a difference in the degradation of DDT under aerobic and anaerobic conditions. DDT degrades more rapidly in soils under anaerobic conditions than in well-aerated soil DDD being the major anaerobic degradation product.[32–34] *Aerobacter aerogenes* has been incubated with

[32] W. D. Guenzi and W. E. Beard, *Science*, 1967, **156**, 1116.
[33] J. F. Parr, G. H. Willis, and S. Smith, *Soil Sci.*, 1970, **110**, 306.
[34] W. D. Burge, *J. Agric. Food Chem.*, 1971, **19**, 375.

DDT in parallel experiments conducted anaerobically and aerobically. In two such cases[35,36] DDD was reported formed under both conditions. Anaerobic conditions favoured the higher yields of DDD. Johnson *et al.*,[37] using pure cultures of micro-organisms, found that no organism, including *A. aerogenes*, produced DDD under aerobic conditions but most produced DDD under anaerobic conditions. Fries[26] has reviewed the literature in this area. He has suggested that authors[37,38] who had shaken their culture flasks to ensure aerobic conditions did not report DDD formation while authors[36,37,39–42] who reported aerobic production of DDD apparently did not shake their cultures. The latter experiments may have involved a period of unintentional anaerobic degradation which could account for the production of much of the reported product. Zoro *et al.*[43] have examined these aspects carefully. They extended earlier suggestions and, working with mixed cultures, including sewage sludge under both anaerobic and aerobic conditions, confirmed that reduced iron–porphyrin complexes effect the reductive dechlorination of DDT in dilute solutions. If, as Zoro *et al.*[43] suggest, this is a major mechanism for the reductive dechlorination of DDT in the environment, it is unlikely to occur under aerobic conditions.

In some studies DDE has been reported as a minor product of DDT degradation (Table 1). The amounts produced were usually less than 10% of the recovered products. In the case of DDE production, at least two mechanisms are possible. One involves chemical dehydrochlorination under alkaline conditions while another involves dehydrochlorinase enzymes in a biological system. Smith and Parr[44] have shown that the threshold pH for the chemical conversion of DDT into DDE in soils is over 12. Since this pH is rarely exceeded in Nature it must be concluded that chemical dehydrochlorination of DDT in the natural environment is uncommon.

The suggestion has been made that DDE may be an intermediate in the formation of DDD. In some cases this is certainly not so. Plimmer *et al.*,[45] using [$^2$H]DDT [2,2-bis-(*p*-chlorophenyl)-2-deuterio-1,1,1-trichloroethane), obtained DDD in which the deuterium atom was retained. Further evidence which suggests that DDD is not formed from DDT *via* DDE is that in most reports of work in which DDE has been incubated with organisms, DDD has

[35] G. Wedemeyer, *Appl. Microbiol.*, 1967, **15**, 569.
[36] J. L. Mendel, A. K. Klein, J. T. Chen, and M. S. Walton, *J. Assoc. Offic. Analyt. Chemists*, 1967, **50**, 897.
[37] B. T. Johnson, R. N. Goodman, and H. S. Goldberg, *Science*, 1967, **157**, 560.
[38] C. I. Chacko, J. L. Lockwood, and M. Zabik, *Science*, 1966, **154**, 893.
[39] K. C. Patil, F. Matsumura, and G. M. Boush, *Appl. Microbiol.*, 1970, **19**, 879.
[40] F. Matsumura and G. M. Boush, *J. Econ. Entomol.*, 1968, **61**, 610.
[41] J. L. Mendel and M. S. Walton, *Science*, 1966, **151**, 1527.
[42] P. S. Barker, F. O. Morrison, and R. S. Whitaker, *Nature*, 1965, **205**, 621.
[43] J. A. Zoro, J. M. Hunter, G. Eglinton, and G. C. Ware, *Nature*, 1974, **247**, 235.
[44] S. Smith and J. F. Parr, *J. Agric. Food Chem.*, 1972, **20**, 839.
[45] J. R. Plimmer, P. C. Kearney, and D. W. von Endt, *J. Agric. Food Chem.*, 1968, **16**, 594.

not been found among the products.[35,45–47] In one case[48] DDE was converted directly into DDD. The mould *Fusarium oxysporum* effected the latter conversion and also converted DDT directly into DDD.

Pathways of further metabolism of DDT degradation products are far from clear. Wedemeyer[35] has proposed the pathway from DDD to DBP in *A. aerogenes* (Figure 1). The pathway was elucidated by incubating proposed intermediates with the organism and examining the products formed in this way. A major criticism of the pathway is that products beyond DDNU could

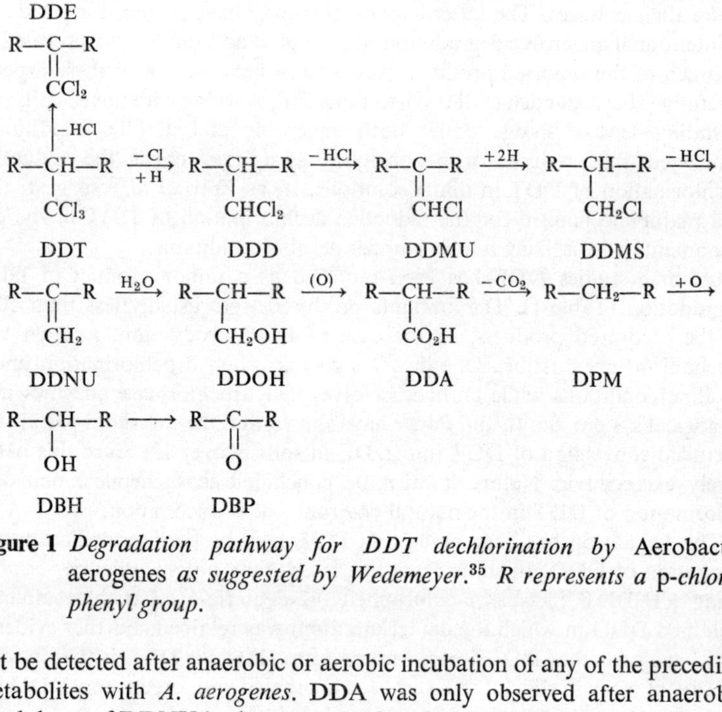


**Figure 1** *Degradation pathway for DDT dechlorination by* Aerobacter aerogenes *as suggested by Wedemeyer.*[35] *R represents a* p-chloro*phenyl group.*

not be detected after anaerobic or aerobic incubation of any of the preceding metabolites with *A. aerogenes*. DDA was only observed after anaerobic breakdown of DDNU in the presence of an exogenous energy source such as glucose. DBP could only be detected by anaerobic breakdown of DDA alone and not if any of the earlier breakdown products were used. This might suggest that earlier breakdown products interfered with the conversion of DDNU into DDA or DDA into DBP. Alternatively, the scheme (Figure 1) may be untenable in its present form and a more complex pattern of breakdown may actually occur.

Fries[26] has closely examined the figures for yields of products of DDA degradation by *A. aerogenes* reported by Wedemeyer[35] (Figure 1). According

[46] G. Wedemeyer, *Science*, 1966, **152,** 647.
[47] B. J. Kallman and A. K. Andrews, *Science*, 1963, **141,** 1050.
[48] R. Engst and M. Kujawa, *Nahrung*, 1967, **11,** 751.

to the scheme (Figure 1), DDA is converted into DBP *via* DPM and DBH. The very low yields of DBH and DBP compared to that of DPM suggest that either DBP is rapidly transformed into some other unknown metabolite or DBH and DBP are only minor products and that the majority of DDA is converted *via* some other pathway into some other undetected product or products. That other pathways of DDT degradation exist is also suggested by some recent studies. When [$^{14}$C]DDT (labelled in the phenyl rings) was incubated with rumen micro-organisms, [$^{14}$C]DDD resulted.[49] Some conversion into polar degradation products also occurred. In this study time-course measurements were made. The amount of [$^{14}$C]DDD increased, reached a maximum, and remained at that level. A similar pattern was followed by the $^{14}$C associated with the polar metabolites, which was in a constant ratio to the [$^{14}$C]DDD. One may conclude that the polar compounds arose *via* a pathway independent of the DDD.

Alexander[50] has isolated many species of marine bacteria which, in pure culture, are effective in converting DDT into water-soluble products. He proposed the use of *Mucor alternans*, a fungus exceptionally active in producing water-soluble products from DDT, as a model for marine bacterial breakdown of DDT. He did not fully characterize products of breakdown of DDT but was able to eliminate DDA, DBH, PCPA (15), and 2-chlorosuccinic acid.

Patil *et al.*[51] have examined isolates of marine micro-organisms for their ability to degrade DDT. They observed that in addition to the predominant metabolite DDD and smaller amounts of DDNS (16), DDOH (17), and DDE, other unknown metabolites were formed.

Recent investigations[52,53] have indicated that DPM, one of the products

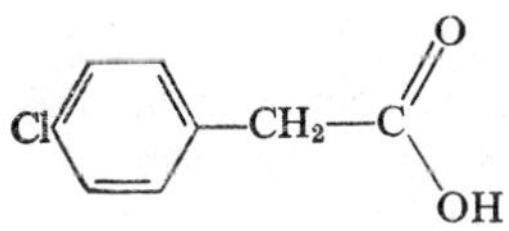


PCPA (15)

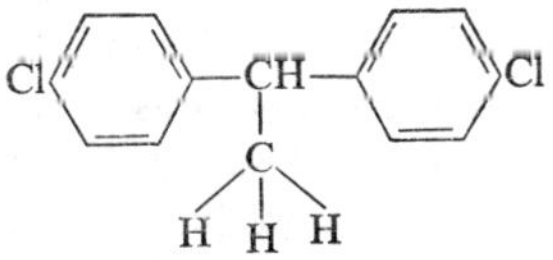


DDNS (16)

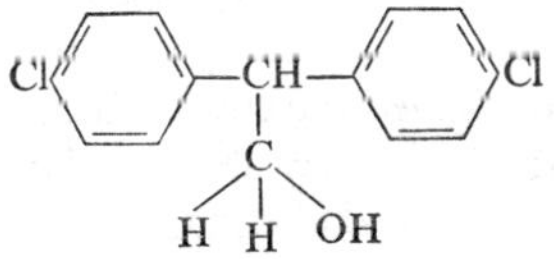


DDOH (17)

[49] G. F. Fries, G. S. Marrow, jun., and C. H. Gordon, *J. Agric. Food Chem.*, 1969, **17**, 860.
[50] M. Alexander, Office of Naval Research, Ann. Rept. No. 1, 1973, National Tech. Information Service, U.S. Dept. Commerce.
[51] K. C. Patil, F. Matsumura, and G. M. Boush, *Environ. Sci. Technol.*, 1972, **6**, 629.
[52] D. D. Focht and M. Alexander, *Science*, 1970, **170**, 91.
[53] D. D. Focht and M. Alexander, *J. Agric. Food Chem.*, 1971, **19**, 20.

of anaerobic breakdown of DDT,[32,35] may be further degraded to a ring-degradation product. Under aerobic conditions a species of *Hydrogenomonas* isolated from sewage converted DPM into PCPA (15). More recently,[54] extracts of *Hydrogenomonas* cells converted DDT anaerobically into DDD, DDMS, DBP, and several other products. When whole cells and oxygen were subsequently added, PCPA (15) was formed, indicating that under appropriate conditions the constituents of a single organism could convert DDT into a ring-degradation product. Focht[55] has described a fungus, *Monilia*, which, in pure culture or mixed culture with *Hydrogenomonas*, converted DPM and PCPA into carbon dioxide, water, and hydrogen chloride. Yields of chloride rarely exceeded 15% and a large proportion of the substrate always remained unchanged.

In conclusion, it may be stated that pathways for the complete biodegradation of DDT in the laboratory have been demonstrated to exist. The pathways appear to be diverse and complex and, although mechanisms for single steps have been described, the complete picture of DDT breakdown is unclear.

Pfaender and Alexander[54] have speculated why DDT continues to persist in Nature in view of its extensive biodegradability in laboratory cultures. One possibility is that complete destruction of DDT requires first anaerobic and subsequently aerobic conditions. Such fluctuations of oxygen may be infrequent in most ecosystems, but do apparently occur in soils. The persistence of DDT in soils suggest that the reason may lie with the actual organisms rather than the prevailing conditions. DDT may be degraded by only relatively few micro-organisms, none of which uses the compound as a carbon or energy source. These micro-organisms thus have no advantage from degrading DDT and must compete with other microbes for available carbon. Their populations thus remain low and degradation of DDT is only effected by very few microbes. Later work by Pfaender and Alexander[56] suggests that a more likely explanation is that DDT persists owing to the fact that cells which can co-metabolize the chemical, although sometimes numerous, do not express a high activity in the environment. Energy in the form of carbonaceous matter must apparently be abundant before cells can express the ability to co-metabolize DDT. In support of this, nutrients added to soil containing DDT speed up the rate of degradation.[33] Pfaender and Alexander[56] have investigated the effect of nutrient addition on the various stages of DDT co-metabolism. They showed that in preparations of sewage organisms, to which DDT was added, DDD, DDE, and DBP were produced anaerobically. The addition of glucose enhanced the rate of DDD formation considerably but slowed DBP production, and the addition of diphenylmethane reduced the rate of formation of both DDD and DBP.

Few reports of microbial degradation of PCB's are available. According to Professor Yoshimura[11] of Kyushu University, Japan, a small part of

[54] F. K. Pfaender and M. Alexander, *J. Agric. Food Chem.*, 1972, **20**, 842.
[55] D. D. Focht, *Bull. Environ. Contam. Toxicol.*, 1972, **7**, 52.
[56] F. K. Pfaender and M. Alexander, *J. Agric. Food Chem.*, 1973, **21**, 397.

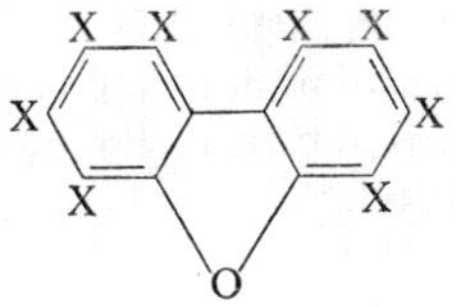


Polychlorinated dibenzofuran (18) (X = Cl or H)

a PCB mixture may be converted into hychoxylated products. Polychlorinated dibenzofurans (18) may also be produced in tissue.[11] PCB's can stimulate the activity of an oxidase system which catalyses the introduction of hydroxy-groups.[11,57–59] A recent report[11] states that laboratory staff of the Enzymes Company of the U.S.A. have discovered a few species of PCB-decomposing microbes among mutant strains created after the microbes, isolated from PCB-contaminated waters and soils, had been irradiated with $\gamma$-rays. Some of the mutant strains were able to decompose up to 95% of PCB in a medium containing 500 p.p.m. of the contaminant. No details of any degradation products were given.

PCB's have been reported[10,60] to be produced from DDT by a mechanism

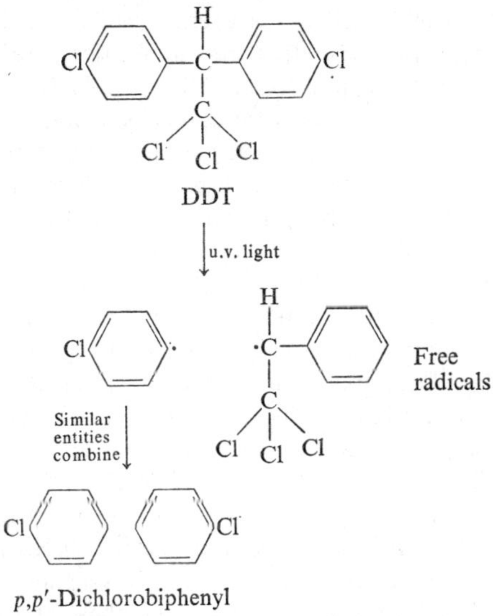


**Figure 2** *A possible mechanism for the conversion of DDT into PCB in the atmosphere or upper layers of the ocean*[10]

57 R. W. Risebrough, P. Reiche, S. G. Herman, D. B. Peakall, and M. N. Kirven, *Nature*, 1968, **220,** 1098.
58 J. L. Lincer and D. B. Peakall, *Nature*, 1970, **228,** 783.
59 J. C. Street, F. M. Urry, D. J. Wagstaff, and A. D. Blau, 'Comparative Effects of Polychlorinated Biphenyls and Organochlorine Pesticides in Induction of Hepatic Microsomal Enzyme,' Presented at ACS meeting, Sept. 8—12, New York, 1969.
60 T. H. Maugh, jun., *Science*, 1973, **180,** 578.

involving free radicals. The mechanism is unlikely to occur in tissue but could be initiated by u.v. radiation in the atmosphere or upper layers of the ocean. A possible reaction is shown in Figure 2,[10] but it could lead to a variety of dichlorobiphenyls only.

## 3 Transport of DDT and PCB's to the Marine Environment

**Transport.**—Chlorinated hydrocarbons, both pesticide and PCB, have become widely distributed over the globe. Two general mechanisms for their transport from their sources may be distinguished, namely by water and by air. For either mechanism the chlorinated hydrocarbons may be in true solution or in vapour form, or adsorbed on waterborne or air-borne particles, respectively. The recent evidence supporting atmospheric transport has been summarized by Risebrough[6] and Goldberg.[5] DDT residues have been detected in rainwater samples. Measurements taken at seven stations in Britain between 1966 and 1967 revealed a mean concentration of 0.08 parts in $10^9$ in the rainwater samples.[61] Residues of DDT in South Florida precipitation averaged 1 part in $10^9$ in eighteen samples taken at four sites between June 1968 and May 1969.[62] Peterle[63] measured DDT residues in meltwater from recent Antarctic snow and found 0.04 parts in $10^9$ of DDT residues. Risebrough *et al.*[64] measured concentrations of chlorinated hydrocarbons (DDT, DDD, DDE) in air-borne dust present in the North Atlantic trade winds at Barbados, West Indies. They assumed 50% efficiency of collection for particles of size greater than 1 $\mu$m and found that the average air concentration of these particles was in excess of $7.8 \times 10^{-14}$ g m$^{-3}$. They speculated that pesticide deposited into the Atlantic from the atmosphere was of a comparable magnitude to that transported to the Atlantic by major rivers. Seba and Prospero[65] measured average air concentrations of DDT and DDE on aerosol particles of size greater than several microns diameter at Barbados two years after the measurements by Risebrough *et al.*[64] By assuming that the collection efficiencies for meshes used were about 50% for particles above several microns in diameter, they recorded levels which averaged $21.2 \times 10^{-14}$ g m$^{-3}$ over a period of 12 days. These authors suggested that the pesticides measured were derived from high altitudes and originated in Europe or North America. Lloyd-Jones[66] pointed out that a significant proportion of DDT and DDE may have been in the vapour phase and consequently the

[61] K. R. Tarrant and J. O'G. Tatton, *Nature*, 1968, **219**, 725.
[62] M. L. Yates, W. Holswade, and A. L. Higer, 1970, quoted in ref. 5.
[63] T. J. Peterle, *Nature*, 1969, **224**, 620.
[64] R. W. Risebrough, R. J. Huggett, J. J. Griffin, and E. D. Goldberg, *Science*, 1968, **159**, 1233.
[65] D. B. Seba and J. M. Prospero, *Atmospheric Environ.*, 1971, **5**, 1043.
[66] C. P. Lloyd-Jones, *Atmospheric Environ.*, 1972, **6**, 283.

figures of Seba and Prospero[65] should be viewed as minimum values for atmospherically transported pesticides.

Residues of DDT in true solution in river waters are generally thought to be in low concentrations. Because of its hydrophobic nature, DDT is readily sorbed on to particulate matter suspended in water. This has been illustrated by Poirrier *et al.*,[67] who showed that colouring colloids (humic material consisting of polymeric hydroxy-carboxylic acids, complexed with varying quantities of iron, of particle size less than 10 $\mu$m) concentrated [$^{14}$C]DDT from natural waters to 15 800 times its original concentration (0.168 parts in $10^9$). Transport of pesticides that are adsorbed on to particulate matter suspended in river waters is well known.[68,69] In a survey of rivers of the western United States over half of the streams where residues were detected contained only 0.005 parts in $10^9$ of DDT residues.[6] Later work showed that residues increased with the silt content of a river and that the maximum recorded level was about 0.1 parts in $10^9$. Much of the pesticide load carried by rivers probably fails to reach the oceans because of deposition of silt and ionic flocculation of dissolved materials from fresh water when saline water is first encountered as it enters an estuary.[69] Some support for this is provided by Seba and Corcoran,[70] who found that the pesticide content of bay water fed by run-off from crop-lands in South Florida was below their detection limit of 0.001 parts in $10^9$. Corcoran and Lee[71] reported that the average concentration of DDT residues in Mississippi Delta river water was only 0.05 parts in $10^9$.

By estimating average river flows and average DDT content in rivers, Risebrough[6] has been able to suggest that about 0.1% of the DDT produced has been carried by surface run-off and rivers into the sea. Estimations involving precipitation and levels of DDT have led to the conclusion that 25% of DDT produced has been carried *via* the atmosphere and rain to the oceans.[6]

Other chlorinated hydrocarbons having similar physical properties to DDT might be transported by similar mechanisms. Dieldrin (19), which is readily volatile,[72] has also been found in rainwater,[61] and in the air, adsorbed on dust.[64] PCB's have been detected in rainwater,[61] in the air,[64,73] and in a 'black snow' which fell on a 10 000 km$^2$ area in South West Sweden in 1969.[74]

Harvey and Miklas[75] have recently analysed for DDT residues and PCB

67 M. A. Poirrier, B. R. Bordelon, and J. L. Laseter, *Environ. Sci. Technol.*, 1972, **6**, 1033.
68 R. Hartung and G. W. Klingler, *Environ. Sci. Technol.*, 1970, **4**, 407.
69 G. E. Walsh, *J. Washington Acad. Sci.*, 1972, **62**, 122.
70 D. B. Seba and E. F. Corcoran, *Pesticide Monitoring J.*, 1969, **3**, 190.
71 E. F. Corcoran and C. C. Lee, Presented at ACS meeting, Dec. 4, New Orleans, 1970.
72 E. P. Lichtenstein, J. P. Anderson, T. W. Fuhremann, and K. R. Schulz, *Science*, 1968, **159**, 1110.
73 S. Jensen, *New Scientist*, 1966, **32**, 612.
74 Smithsonian Institution Centre for Short-Lived Phenomena, 1970, Event 19-70, Item 876.
75 G. R. Harvey and H. P. Miklas, 'A Study Program to Identify Problems Related to Oceanic Environmental Quality', International Decade of Ocean Exploration Report, National Science Foundation (U.S.), 1972, p. 469.

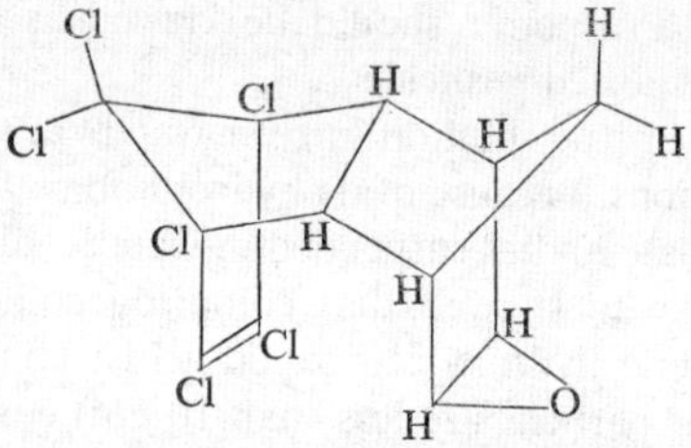


(19)

in phytoplankton across the North and South Atlantic ocean. They concluded, for a number of reasons, that the atmosphere provides the major path for delivery of both PCB and DDT to the oceans. Their reasons may be summarized as:

(1) There was a uniformity of chlorinated hydrocarbon concentrations in plankton across the North Atlantic consistent with these compounds being added directly from the atmosphere.

(2) There was little difference in the average concentrations of these pollutants between the North and South Atlantic. Considering the different levels of industrial activity and lack of North to South ocean circulation, atmospheric transport from Europe and North America to the Southern Hemisphere seems the most likely explanation. Transport of this kind has been shown by atmospheric movement of various tracers, including radioactive fallout.

(3) The high concentration of PCB in a plankton sample collected off Iceland in the East Greenland current, bringing water from the Arctic, could only be explained in terms of atmospheric transport, since ocean circulation is generally slow. Evidence was provided which suggested that DDT and PCB behave differently in the atmosphere and in the oceans. Data for the South Atlantic showed a considerable east to west gradient of concentration in plankton for both types of chlorinated hydrocarbon. The gradient for PCB was, however, more than thirty-fold while for DDT it was little more than ten. The gradient appears explicable only by the exhaustion of the pollutants from the air mass during passage over the ocean in the South Atlantic westerlies.

**Sewage Sludge.**—A less obvious source of chlorinated hydrocarbons in the marine environment is sewage sludge. Because of its high organic content, sludge would be expected to concentrate hydrophobic pollutants like chlorinated hydrocarbons. In 1970, Holden[76] found high residues of PCB in fish and molluscs from waters of the Firth of Clyde. He traced the source to Glasgow's sewage sludge, which had been dumped in the estuary. An estimated 1 ton per year of PCB entered the estuary from this source. In a

[76] A. V. Holden, *Nature*, 1970, **228**, 1220.

similar way PCB was shown to emanate from sewage plants in Manchester and London. It proved impossible to trace the specific sources of PCB's because several factories were linked to the sewers. Appreciable amounts of PCB and DDT residues are reportedly added to Californian coastal waters from urban sewage outfalls.[77] An estimated 50 tons per year of PCB and a similar amount of DDT residues came from a particularly polluted discharge point serving a highly industrial area.

## 4 Distribution of DDT and PCB's in the Marine Environment

In the marine environment, pesticide residues and PCB's may be physically distributed in different ways. They may be in true solution in sea-water, associated with both inorganic and organic particulate matter suspended in ocean water, in sediments, in surface slicks, or within organisms.

**Sea-water.**—DDT residues have been measured in sea-water by Cox.[78] He found concentrations in whole sea-water, determined by continuous-flow liquid–liquid extraction, ranging from 0.0023 parts in $10^9$ for water off Oregon and Washington to 0.0056 parts in $10^9$ for waters off Southern California. Evidence was provided by Cox that particles of diameter less than 2 $\mu$m carry most of the DDT residues in the sea. It may be assumed that much of the particulate matter is eventually circulated to the deep ocean or is deposited in sediments. Where values are available,[11,79,80] the concentrations of chlorinated hydrocarbons in sediments are generally higher than those found in overlying water. Laboratory samples of water and sediment from the Fraser River, and Georgia Strait, British Columbia, were treated with DDT and DDD in a recent study.[81] After six weeks incubation at 13 °C all detectable DDT and DDD had moved into the sediment. In a review on pesticide–sediment–water interactions, Pionke and Chesters[82] considered the factors which affect pesticide adsorption from water on to organic matter, clay particles, and $Fe^{2+}$ and $Fe^{3+}$ gels and which affect the subsequent sedimentation. By comparing adsorption isotherms for pesticides on soil particles and extending such behaviour to particles in aquatic sediments, these authors concluded that chlorinated hydrocarbons should be more strongly adsorbed on to sediments than on to soils, and they should tend to concentrate therefore in sediments.

PCB's were found in high concentrations in sediments near centres of high industrial activity.[11,12,76,77] Residues in the industrially polluted Escambia Bay, U.S.A., increased from water to sediment.[79] The water contained a

[77] T. T. Schmidt, R. W. Risebrough, and F. Gress, *Bull. Environ. Contam. Toxicol.*, 1971, **6**, 235.
[78] J. L. Cox, *Fisheries Bull.*, 1971, **69**, 443.
[79] T. W. Duke, J. I. Lowe, and A. J. Wilson, *Bull. Environ. Contam. Toxicol.*, 1970, **5**, 171.
[80] J. R. W. Miles and C. R. Harris, *Pesticide Monitoring J.*, 1973, **6**, 363.
[81] P. C. Oloffs, L. J. Albright, S. Y. Szeto, and J. Lau, *J. Fisheries Res. Board Canada*, 1973, **30**, 1619.
[82] H. B. Pionke and G. Chesters, *J. Environ. Quality*, 1973, **2**, 29.

maximum concentration of 275 parts in $10^9$ and sediment a maximum of 486 parts in $10^9$.

**Sea Surface.**—The interface between the sea and atmosphere is a unique environment. Hydrophobic compounds such as hydrocarbons, long-chain fatty acids and alcohols[83,84] and chlorinated hydrocarbons[70,85] gather at the surface. Some authors[65] have suggested that airborne particles carrying pesticide residues, on entry into the marine environment, will reside briefly at the sea surface, especially if the particles are not easily wetted. Under suitable wind and sea conditions the sea surface may be compressed into a visible slick.[70] Surface slicks, having abundant nutrients and organic matter, are areas of high biological activity.[86] It is likely that pesticides associated with surface layers could easily enter the biosphere.

**Organisms.**—DDT was first recognized as an environmental contaminant by its occurrence in organisms. Its recognition was made possible because of the ability of organisms to concentrate lipophilic compounds to levels within the analytical capabilities of the available instrumentation. DDT residues have been analysed from most kinds of organism and foodstuffs. Levels vary widely from species to species and within species but sufficient work has been done to arrive at average values for the important species (plankton, fish, *etc.*) in the marine environment.

Edwards[87] has reviewed the literature reporting studies of DDT in non-target aquatic communities (those for which the pesticide was not intended) and has concluded that phytoplankton have a high capacity for accumulation of DDT. Cox[14] has reviewed the occurrence of DDT residues in marine phytoplankton. He concludes, from his own determinations, that values for DDT residues in phytoplankton range from 1.2 to 2.7 $\mu$g [g organic carbon (dry weight)]$^{-1}$. He pointed out that the method of collection would not exclude non-living particulate matter. Harvey *et al.*[88] reported the range 0.2—0.5 ng [g (wet weight)]$^{-1}$ for DDT residues in Sargassum weed. The average concentration of DDT residues for all marine phytoplankton has been put at 0.01 p.p.m. by Goldberg[5] (Table 2).

Levels of DDT residues in marine fish are presented in reviews by Johnson,[89] Risebrough,[6] Otterlind *et al.*,[90] and Walsh.[69] Values vary (Table 2), but Goldberg[5] has suggested an approximate average of 1 p.p.m. for all fish.

Less is known of the levels of pesticides in marine mammals. Grey whales

[83] W. D. Garret, *Deep-Sea Res.*, 1967, **14**, 221.
[84] P. M. Williams, *Deep-Sea Res.*, 1967, **14**, 791.
[85] R. A. Duce, J. G. Quinn, C. E. Olney, S. R. Piotrowicz, B. J. Ray, and T. L. Wade, *Science*, 1972, **176**, 161.
[86] W. H. Sutcliffe, jun., E. R. Baylor, and D. W. Menzel, *Deep-Sea Res.*, 1963, **10**, 233.
[87] C. A. Edwards, *Crit. Rev. Environ. Contr.*, 1970, **1**, 7.
[88] G. R. Harvey, J. T. Bowen, R. H. Backus, and G. D. Grice, 1971, quoted in ref. 14.
[89] D. W. Johnson, *Trans. Amer. Fisheries Soc.*, 1968, **97**, 389.
[90] G. Otterlind, S. Jensen, and M. Olsson, International Council for the Exploration of the Sea, C.M. 1971/E.31.

**Table 2** *Examples of the range of values of DDT residues (DDT, DDE, DDD) and PCB in some environments and organisms*

| | *Range of values* | | |
|---|---|---|---|
| *Environment* | *DDT: Mass/* µg kg$^{-1}$ | *PCB: Mass/* µg kg$^{-1}$ | *Reference* |
| Rain-water | 0.08* | | 61 |
| Air-borne dust | up to 50 | up to 5 | 64 |
| Silt-free rivers | 0.001—0.005 | | |
| Silty rivers | 0.005—0.12 | | 6 |
| Lake water (Michigan) | 0.001—0.003 | | |
| Sea-water (including particulate matter) | 0.0023—0.1 | | 78 |
| Highly polluted harbour (Port of Tagonoura, Japan) | | | |
| water | | 16.8—23.6 | 11 |
| sediment | | 360—470 | |
| Sediment (off Western U.S.) | 0.04—1.5 | 0.3—1.8 | 115 |
| *Organism* | *Mass*/µg g$^{-1}$ | *Mass*/µg g$^{-1}$ | |
| Plankton (wet weight) | 0.0002—0.0026 | 0.007—0.64 | 75 |
| Plankton (lipids) | | 100—900 | |
| Plankton | 0.01* | | 5 |
| Invertebrates | | | |
| various (wet wt.) | | 0.5—2.0 | 12 |
| pink shrimp (wet wt.) | 0.002—0.005 | 0.01—0.07 | 115 |
| Fish (whole) | 0.02—14.0 | 0.02—1.2 | 112, 6 |
| Marine mammals (blubber) | 0.4—800 | | 91, 93 |
| seal (fat from tail) | | 30—200 | 12 |
| Birds | | | |
| (whole body) | 3.0—66 | 0.35—24 | 6 |
| (body fat) | 41—199 | 53—697 | |

* Average value

(*Escherichtius robustus*) feed largely on benthic organisms while sperm whales (*Physeter catodon*) feed on larger pelagic organisms. The blubber of the former contains 0.4 p.p.m. and that of the latter up to 6 p.p.m. of DDT residues.[91] Anas and Wilson[92] examined fur-seal pups for pesticide residues. They found a maximum value of 45 p.p.m. of DDE in one pup and all pups contained DDD, while most contained DDT. Butler[93] reported up to 800 p.p.m. of DDT residues in porpoise blubber.

DDE almost always comprises at least 80% of the DDT residues found in marine organisms. The remainder consists principally of *p,p'*-DDT and *p,p'*-DDD.[6]

[91] A. A. Wolman and A. J. Wilson, jun., *Pesticide Monitoring J.*, 1970, **4**, 8.
[92] R. E. Anas and A. J. Wilson, jun., *Pesticide Monitoring J.*, 1970, **3**, 198.
[93] P. A. Butler, in Proceedings of the Marsh and Estuary Management Symposium, Baton Rouge, Louisiana, ed. J. D. Newson, T. J. Moran's Sons Inc., Louisiana, 1968.

The literature which reports levels of PCB in marine organisms has recently been reviewed.[5,6,10–12,94,95] Tanaka[11] suggested that the highest levels of PCB are recorded for fat-rich marine animals, *e.g.* seals (3—312 p.p.m. of their fat), followed in order of decreasing levels by birds that feed on fish (6.5 p.p.m. in sea gulls), fishes, invertebrates, and plants. The PCB concentrations in fish varied markedly with their habitat: those from highly polluted inland seas showed higher values (*e.g.* 1—1.2 p.p.m. in *Engraulis japonica* and sea bass) than those from open-sea waters (*e.g.* 0—0.04 p.p.m. in tuna and bonito). Tatsukawa and Isono[12] presented figures which indicated that amounts of PCB found in birds and seals are often many hundreds of times those found in fish and invertebrates (Table 2). Fish from Japanese waters contain amounts of PCB comparable with those of birds from other countries. This is the result of the more highly contaminated marine environment around Japan.[12]

## 5 Uptake of Pesticide Residues by Organisms

**Laboratory Studies.**—In the laboratory an organism constantly exposed to pesticide will build up concentrations of the pesticide in amounts dependent on the duration and degree of exposure. A plateau concentration may eventually be reached, suggesting that uptake is balanced by excretion and/or metabolism of the pesticide. When exposure continues for long enough, the level of pesticide in the tissue may decline again, despite continued constant exposure. In other cases, concentrations have increased significantly after a further period of exposure to the pesticide.[96]

Experimental data from fish are in agreement with these ideas. Reticulate sculpins (*Cottus perplexus*) were exposed to six concentrations of HEOD [dieldrin, (19)] from 17 to 860 parts in $10^9$.[97] Whole fish were analysed at intervals. Plateau concentrations of HEOD in fish were approached but exposure was not continued long enough to see if decline or further increase occurred. Addition of any of the fish contaminated with HEOD to clean water resulted in an exponential decline of HEOD concentration in the fish. Earlier literature indicates that chlorinated hydrocarbons reach plateau levels in fish.[98]

In the longer term, fish inhabiting natural waters containing DDT residues appear to accumulate amounts dependent on their total exposure. Youngs *et al.*[99] found a significant correlation between the age of lake trout (*Salvelinus*

[94] E. H. Dustman, L. F. Stickel, L. J. Bus, W. L. Reichel, and S. N. Wiemeyer, Transactions of the 36th Conference on North American Wildlife and Natural Resources, Portland, Oregon, 7—10 Mar. 1971, pp. 118—133, Washington D.C. Wildlife Management Institute, 1971.
[95] J. E. Portmann, International Council for the Exploration of the Sea, C.M. 1970/E.9.
[96] F. Moriarty, *Sci. Total Environ.*, 1972, **1**, 267.
[97] G. G. Chadwick and R. W. Brocksen, *J. Wildlife Management*, 1969, **33**, 693.
[98] O. B. Cope, *Ann. Rev. Entomol.*, 1971, **16**, 325.
[99] W. G. Youngs, W. H. Gutenman, and D. J. Lisk, *J. Environ. Sci. Technol.*, 1972, **6**, 451.

*namaycush*) and the total body residues of DDT compounds, which ranged from 1 p.p.m. at one year to 28 p.p.m. at eleven to twelve years of age.

Sodergren[100] investigated the uptake of [$^{14}$C]DDT by a freshwater alga (*Chlorella* sp.) and concluded that:

(*a*) uptake was rapid (of the order of minutes);

(*b*) incorporated DDT was not released when the algae were added to DDT-free water;

(*c*) uptake was quantitative.

In a later study these results were confirmed by the same author.[100a] Cox[14] has suggested two mechanisms for the uptake of DDT by phytoplankton. One mechanism involves phase partitioning of DDT between sea-water and the lipid portion of algal cells. The other involves adsorption of DDT on to cell surfaces. The latter mechanism was proposed after it was discovered that a saturation value of [$^{14}$C]DDT, accumulated by *Dunaliella salina*, was independent of the ambient concentration. Cox[78] provided evidence that in the ocean the majority of DDT is adsorbed on to particles of less than 1—2 $\mu$m diameter and that the DDT would in consequence be largely unavailable for uptake by phytoplankton.

Some species of aquatic organisms show a greater capacity for accumulation of pesticide residues than do others. An example is provided by spot and croaker fish which, when exposed to DDT at a concentration of 0.1 parts in $10^9$, had whole-body concentrations ranging from 1 to 3.8 p.p.m. of DDT.[101] Pinfish exposed to the same concentration of DDT stored 2.4 times as much DDT as did croaker fish.

The degree of accumulation by an organism also depends on the pesticide to which the organism is exposed. Snails (*Physa* sp.) concentrated DDT to a level 35 000 times greater than the ambient concentration. This was less than the concentration of ethoxychlor (98 000 times) and methoxychlor (120 000 times).[102] When three species of unicellular algae and two species of ciliates were exposed to pesticides (DDT and parathion), each at a concentration of 1 p.p.m., for seven days, the organisms concentrated DDT to 99—964 p.p.m. and parathion to 50—116 p.p.m.[103]

Values of plateau concentrations for one insecticide in an organism may be affected several fold by subsequent exposure to a second insecticide,[104] and simultaneous exposure to two insecticides can also affect the plateau concentrations of both insecticides.[105,106]

**Biological Magnification.**—'Biological magnification' and 'ecological magnification' are terms used to describe the increasing abundance of a pollutant

[100] A. Sodergren, *Oikos*, 1968, **19**, 126.
[100a] A. Sodergren, *Oikos*, 1971, **22**, 215.
[101] D. J. Hansen and A. J. Wilson, jun., *Pesticide Monitoring J.*, 1970, **4**, 51.
[102] I. P. Kapoor, R. L. Metcalf, A. S. Hirwe, P.-Y. Lu, J. R. Coats, and R. F. Nystrom, *J. Agric. Food Chem.*, 1972, **20**, 1.
[103] W. W. Gregory, jun., J. K. Reed, and L. E. Priester, jun., *J. Protozool.*, 1969, **16**, 69.
[104] W. B. Deichmann, and W. E. MacDonald, *Food Cosmet. Toxicol.*, 1971, **9**, 91.
[105] W. B. Deichmann, W. E. MacDonald, and D. A. Cubit, *Science*, 1971, **172**, 275.
[106] D. J. Wagstaff and J. C. Street, *Bull. Environ. Contam. Toxicol.*, 1971, **6**, 273.

in animals from lower to higher trophic levels in a food web. Animals of different species have varied diets, but all depend ultimately on plants for their energy. Herbivares are usually preyed on by carnivores, which in turn may be preyed on by higher carnivores, and so on. There is, however, little evidence that links the amounts of pesticides within animals to their position in a food web.[96] That there should be concentration in food webs implies that the major source of pesticide is the food, not the environment. This appears to be untrue for aquatic organisms.

Chadwick and Brocksen[97] studied the uptake of HEOD (19) from contaminated water and food by the reticulate sculpin fish (*Cottus perplexus*). The fish were held in water containing 0.5 parts in $10^9$ of HEOD. One group were fed tubificid worms containing known amounts of HEOD while another group were fed uncontaminated worms. There was no difference in the total amount of the chlorinated hydrocarbon accumulated by either group of fish. Obviously the fish absorbed their HEOD directly from the water. Reinert[107] described similar results for a food web consisting of alga, crustacean, and fish, exposed to HEOD. By contrast, Macek and Korn[108] showed in separate experiments that brook trout (*Salvelinus fontinalis*) exposed to [$^{14}$C]DDT accumulated approximately ten times more of the available DDT from food than directly from the water. They concluded that since the concentration of DDT is much higher in food than in water, the food web is the major source of DDT for fish in natural waters. These authors assumed, possibly erroneously, that the intakes from food and water were additive, whereas Chadwick and Brocksen[97] have showed for HEOD that there can be an interaction, with one source affecting the intake from the other source.

Larger scale field experiments reinforce the conclusion that for aquatic organisms the food-chain is unimportant. Meeks[109] sprayed a four-acre marsh with DDT at 0.2 lb acre$^{-1}$. The DDT was labelled with $^{36}$Cl in the phenyl groups, and DDT concentrations in samples of water, sediment, and food-chain organisms were estimated from the radioactivity present. The concentration of DDT in the water had almost reached saturation within 12 hours of spraying, and DDT had disappeared within one month. Particulate matter suspended in the water lost all activity within one week and surface sediment concentrations of DDT remained constant at 0.3 p.p.m. from six weeks onwards. Plants, invertebrates, and fish all reached near maximum values of DDT contamination within a week and had much higher concentrations than the water. Patterns of accumulation, then loss of DDT, were consistent for all species examined. Magnification of DDT along the food web would undoubtedly have resulted in successive trophic levels reaching maximum values of DDT one after another. The data appear consistent with

[107] R. E. Reinert, Ph.D. Dissertation, University of Michigan, Ann Arbor, 1967, quoted in ref. 96.
[108] K. J. Macek and S. Korn, *J. Fisheries Res. Board Canada*, 1970, **27**, 1496.
[109] R. L. Meeks, *J. Wildlife Management*, 1968, **32**, 376.

direct absorption of DDT from the water by all species. This conclusion seems to be at some variance with other field data.

**Field Studies.**—Studies on naturally occurring food-chains have revealed that animals in higher trophic levels have higher concentrations of insecticide residues[79,93,110,111] and PCB's[12,112] than animals of each preceding trophic level. Moriarty[96] has reviewed the reports of biological magnification and has suggested a possible reason. Predators do not take random samples of prey species. They are perhaps more likely to catch organisms which, because of higher levels of insecticide than other members of the species, find it less easy to avoid capture. Cooke[113] has provided an elegant demonstration of this. Newts (*Triturus cristatus*) were offered pairs of frog tadpoles (*Rana temporaria*) as food. One of each pair was first intoxicated by exposure to DDT, the other was a normal control. On 90 occasions out of 100 the newts chose the contaminated tadpole. Moriarty[96] concluded that where terrestrial predators have higher concentrations of pesticide than random samples of their prey species, they may have selectively fed on the most highly contaminated individuals of the prey. Marine organisms also may be subjected to widely differing concentration of chlorinated hydrocarbons in their environment. Concentrations vary widely between the main body of sea-water, the surface microlayer, and particulate matter. Such variations could significantly affect the extent of accumulation of chlorinated hydrocarbons by different species.

Long-term projections of the future distribution of DDT for an inland lake environment were computed by using mathematical models. Harrison *et al.*[114] predicted that DDT concentrations in long-lived species in higher trophic levels may be expected to continue long after the addition of DDT to the environment has ceased. The model they used assumed that the food web was the dominant means of magnification of DDT in organisms. Available evidence suggests that, for aquatic species at least, the food web is irrelevant and that DDT is taken up directly from the physical environment. Amounts in individual species start to decrease when the source of contamination is removed.

## 6 Analysis of Chlorinated Hydrocarbons

An abbreviated procedure for the analysis of chlorinated hydrocarbons from a sample organism is illustrated in Figure 3. More detailed procedures for the analysis of chlorinated hydrocarbons may be found elsewhere.[75,115]

110 G. M. Woodwell, C. F. Wurster, and P. A. Isaacson, *Science*, 1967, **156**, 821.
111 E. H. Dustman and L. F. Stickel, *Ann. N. Y. Acad. Sci.*, 1969, **160**, 162.
112 S. Jensen, A. G. Johnels, S. Olsson, and G. Otterlind, *Nature*, 1969, **224**, 247.
113 A. S. Cooke, *Nature*, 1971, **229**, 275.
114 H. L. Harrison, O. G. Loucks, J. W. Mitchell, D. F. Parkhurst, C. R. Tracy, D. G. Watts, and V. J. Yannacone, *Science*, 1970, **170**, 503.
115 R. R. Claeys, International Decade of Ocean Exploration Report, National Science Foundation, U.S., 1972, p. 449.

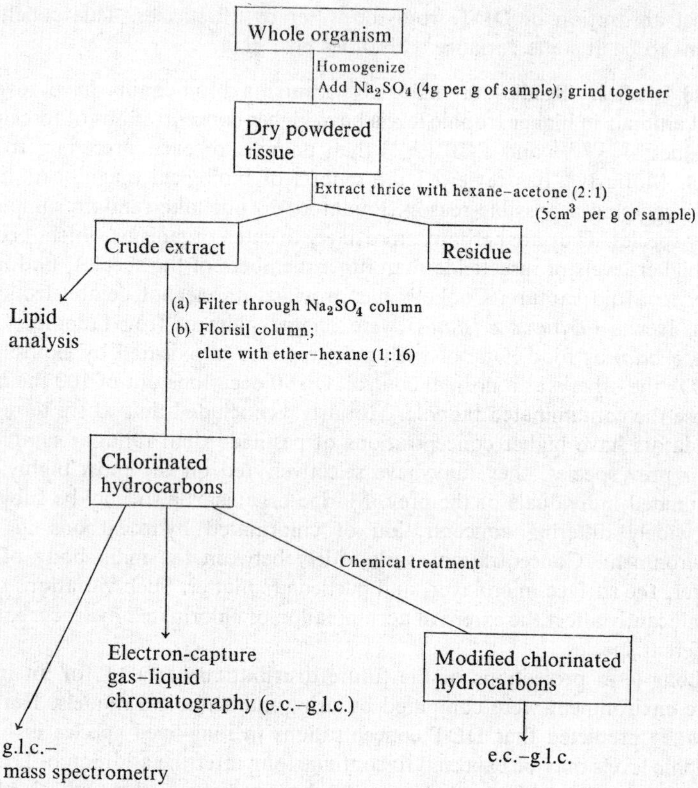


**Figure 3** *A simplified scheme for the analysis of chlorinated hydrocarbons from an organism*

A typical analysis generally involves three basic steps. These are: (i) extraction of chlorinated hydrocarbons (together with lipids) using a suitable solvent mixture; (ii) removal of lipids from the mixture using a Florisil column; and (iii) final analysis. An aliquot is usually removed prior to Florisil clean-up for a lipid analysis. This enables the quantitative assessment of chlorinated hydrocarbons to be related to the fat content of the organism and also to the whole organism.

Final analysis involves injection of the chlorinated hydrocarbon mixture into a gas–liquid chromatograph equipped with an electron-capture sensitive detector. Under ideal conditions a peak on a chromatographic trace may be identified after co-injection of the mixture with an appropriate standard compound. Individual components of a mixture of chlorinated hydrocarbons have widely differing sensitivities to an electron-capture detector. Quantitative assessment of a component in a mixture requires comparison of its peak height with the peak heights of the appropriate standard after a series of calibration injections. A major difficulty encountered in g.l.c. analysis of tissue is that very many peaks (representing DDT residues, PCB's, and other pesticides) tend to crowd together and often coincide. Various attempts have been made to overcome the difficulties and to improve the analysis of chlorinated hydrocarbons.

Interpretation of g.l.c. analysis may be facilitated by comparing chromatographic traces after a series of injections of the chlorinated hydrocarbon mixture through different g.l.c. columns. Individual components of the mixture may behave differently on different columns, and as a result the relative positions of peaks may alter.

A more certain, but expensive, method of analysis involves linking the g.l.c. to a mass spectrometer. As components emerge from the chromatographic column they pass into a mass spectrometer which can produce characteristic spectra for components of the mixture, and which may reveal more than one compound from coincident peaks.

Mixtures of chlorinated hydrocarbons may be simplified by chemical treatment before the g.l.c. stage. Methods that have been used are dehydrochlorination, nitration, and perchlorination. Treatment of a mixture of chlorinated hydrocarbons with strong alkali would convert, for example, *p,p'*-DDT into the dehydrochlorinated product, *p,p'*-DDE. The latter has a shorter retention time in the g.l.c. than *p,p'*-DDT and also is more sensitive than *p,p'*-DDT to electron-capture detection. Treatment of a mixture of PCB and DDT residues with a mixture of concentrated sulphuric and nitric acids has the effect of nitrating the DDT residues while the PCB's remain unaffected. Perchlorination of PCB produces a single highly-chlorinated (and therefore highly sensitive to electron-capture detection) compound. Chemical treatment of chlorinated hydrocarbon mixtures has been discussed by Veith and Lee.[116]

The limit of detection of an electron-capture detector for the chlorinated hydrocarbons (*e.g. p,p'*-DDE) to which it is most sensitive is approximately $10^{-13}$ g. At levels approaching this, a major problem is the avoidance of contamination. Solvents used for analysis of chlorinated hydrocarbons have to be specially purified, glassware carefully cleaned, and materials like plastics, paper, and mineral oil strictly avoided.

[116] G. D. Veith and G. F. Lee, *Water Res.*, 1970, **4**, 265.

# 7

# Environmental Organic Chemistry of 2,4-Dichlorophenoxyacetic Acid

BY J. E. ALLEBONE, R. J. HAMILTON, AND B. RAVENSCROFT

## 1 Introduction

Although there are many reviews and books dealing with plant growth-regulators,[1—7] it was felt that the first Specialist Report in this series on 2,4-dichlorophenoxyacetic acid (2,4-D) should include some references to work outside the 1971—1973 period covered by the Volume. From an industrial point of view, 2,4-D is so cheap to manufacture that little research can be sponsored from the profit margins on phenoxyacetic acid herbicides. Despite this, there is a very large volume of work being performed by Government agencies.

Chemical weedkilling has been practised seriously for about seventy years but it was not until after the second World War that herbicides were used on a large scale. The discovery and use of plant growth-regulators which can kill weeds without affecting corn has increased the yield of grain crops and, arguably, has done more to improve the standard of living than any other single chemical. In 1940, Dr. William Templeman[8] noted the selective weed-killing action of the phenoxyacetic acids, man-made plant growth-regulators. The chlorophenoxy herbicides (Table 1) kill plants by contact action and also by being translocated from the tops into the roots. They are selective in killing many broad-leaved weeds in the presence of cereal and grass crops. They can also be used as pre-emergence soil sterilants[9] and as pre-emergence herbicides[10] to be absorbed from the soil by young seedlings.

1 'Chemistry and Uses of Pesticides', E. R. de Ong, Chapman and Hall, London, 1956.
2 'Physiology and Biochemistry of Herbicides', ed. L. J. Audus, Academic Press, London, 1960.
3 'Chemistry and Mode of Action of Herbicides', A. S. Crafts, Interscience, New York, 1961.
4 R. W. Bovey, *Economic Botany* 1971, **25**, 385.
5 W. R. Benson and C. R. Blalock, *J. Assoc. Offic. Agric. Chemists*, 1971, **54**, 192.
6 M. M. Robertson and R. C. Kirkwood, *Weed Res.*, 1970, **10**, 94.
7 M. A. Loos in 'Degradation of Herbicides', ed. P. C. Kearney and D. D. Kaufman, Marcel Dekker, New York, 1969.
8 R. E. Slade, W. G. Templeman, and W. A. Sexton, *Nature*, 1945, **155**, 497.
9 D. Bohmont, *Western Weed Control Conf. Research Progress Rept.*, 1952, **13**, 12.
10 J. C. Anderson and G. Ahlgren, *Down to Earth*, 1947, **3**, 16.

**Table 1** *Phenoxyacetic acids*

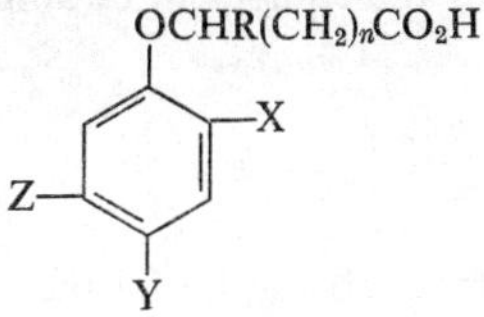


| Common name | Full name | R | X | Y | Z | *n* | *m.p.*/°C | *Solubility in water/ p.p.m.* |
|---|---|---|---|---|---|---|---|---|
| 2,4-D | 2,4-dichlorophenoxy-acetic acid | H | Cl | Cl | H | 0 | 140.5 | 620 |
| MCPA | 2-methyl-4-chloro-phenoxyacetic acid | H | Me | Cl | H | 0 | 119.0 | 825 |
| 2,4,5-T | 2,4,5-trichloro-phenoxyacetic acid | H | Cl | Cl | Cl | 0 | 158.0 | 251 |
| MCPP, mecoprop | 2-(2-methyl-4-chloro-phenoxy)acetic acid | Me | Me | Cl | H | 0 | 94.0—95.0 | 620 |
| 2,4-DP, dichlorprop | 2-(2,4-dichloro-phenoxy)propionic acid | Me | Cl | Cl | H | 0 | 117.5—118.1 | 350 |
| silvex | 2-(2,4,5-trichloro-phenoxy)propionic acid | Me | Cl | Cl | Cl | 0 | 179.0—181.0 | 140 |
| MCPB | 4-(2-methyl-4-chloro-phenoxy)butyric acid | H | Me | Cl | H | 2 | 100.0 | 44 |
| 2,4-DB | 4-(2,4-dichloro-phenoxy)butyric acid | H | Cl | Cl | H | 2 | 117.0—119.0 | 40 |
| 2,4,5-TB | 4-(2,4,5-trichloro-phenoxy)butyric acid | H | Cl | Cl | Cl | 2 | 114.0—115.0 | — |

By 1947, it was realised that large-volume application was not necessary and that low-volume application lowered the cost. The use increased from treating 100 000 acres in 1946 to several million in 1947. Phenoxyacetic acid derivatives were cheap to use, costing 25p per acre in those days. Although newer growth-regulating herbicides have been discovered and applied, phenoxyacetic acids are still very important, *e.g.* during 1966 in the Rhineland it is estimated[11] that 62% of the growth-regulating herbicides applied to cereals contained phenoxyacetic acids. Their most infamous use has been in Vietnam, where more than 2 million hectares of woodland (20% of the South Vietnam forests) have been defoliated by using phenoxyacetic acids at a rate of 29 kg $ha^{-1}$ (*i.e.* a total of $>4 \times 10^7$ kg). It is estimated[12] that it will take at least a decade for the forests to be regenerated. The land will recover

[11] R. Jacob-Haupt, *Gesunde Pfl.*, 1967, **19**, 140.
[12] A. H. Westing, *Bioscience*, 1971, **21**, 893.

because the phenoxyacetic herbicides are not very persistent in soil, an obvious advantage in crop rotation. They are still cheap, a typical ester formulation costing £8.00 per gallon, or the equivalent of 50p per acre of cereal crops or £1 per acre of grassland.

## 2 Synthesis

Few changes have been made to the initial methods of synthesis shown in Scheme 1.[13,14] Only recently have studies been made of the contaminants

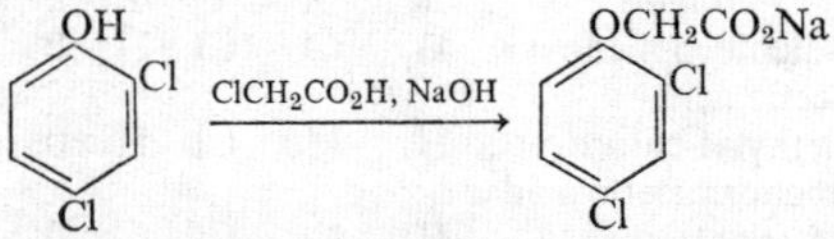


**Scheme 1**

which are present in technical grade 2,4-D. Three contaminants, bis-2,6-dichlorophenoxymethane (1), 2,2′,4,6′-tetrachlorodiphenoxymethane (2), and bis-2,4-dichlorophenoxymethane (3), have been shown to have no

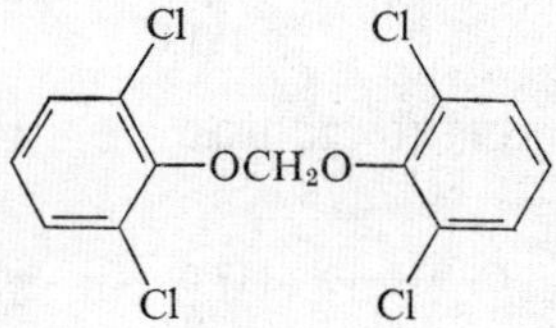


(1)

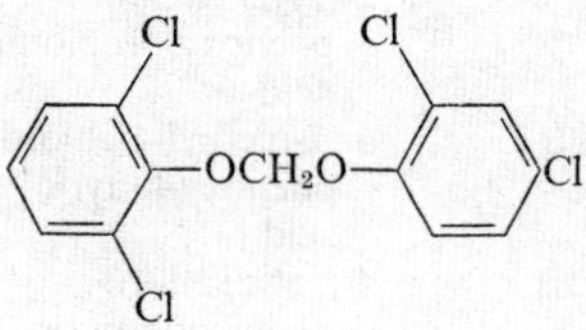


(2)

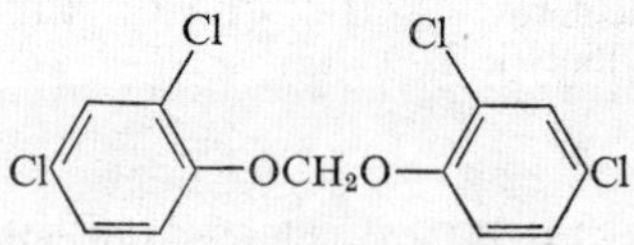


(3)

teratogenic effects on mammals.[15] The annual production of 2,4-D in Great Britain and the United States of America is estimated to be five thousand and twenty thousand tons, respectively.

## 3 Distribution of 2,4-D in the Environment

**Plants.**—The visible effects of 2,4-D are apparent within a few days in sensitive plants. The herbicide acts like an indole auxin which has an unusually

[13] J. K. Plimmer, 'Kirk Othmer Encyclopaedia of Chemical Technology,' 2nd Edn., Vol. 22, 1970, p. 180.
[14] R. Pokorny, *J. Amer. Chem. Soc.*, 1941, **63,** 1768.
[15] B. L. Huston, *J. Agric. Food. Chem.*, 1972, **20,** 724.

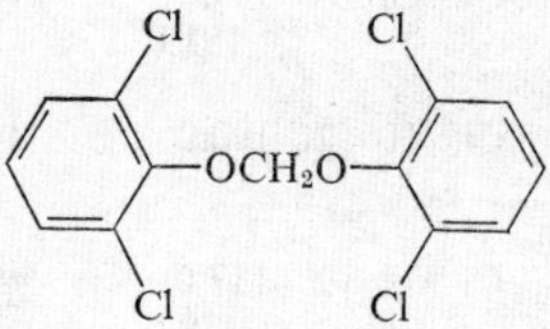

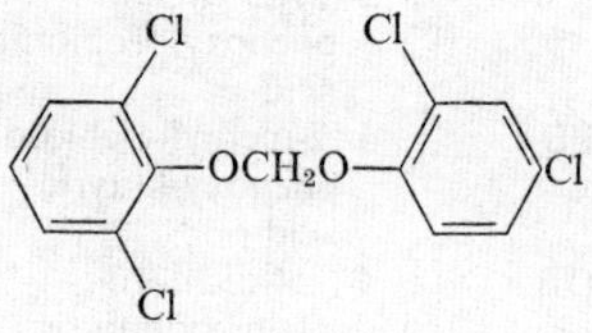

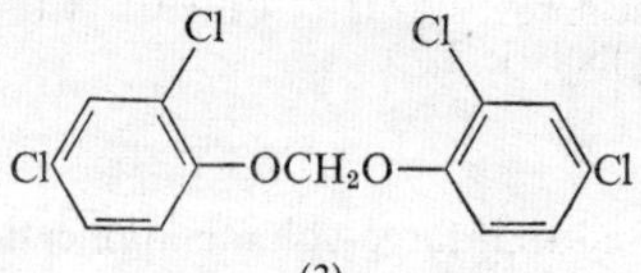

long persistence, causing cell division[16] and cell elongation,[17] resulting in abnormalities in leaf shape,[18—20] in increased lateral root growth, in fusion of flower parts and disruption of pollen release,[21] and in the promotion of fruit growth.[22] It also inhibits seed germination,[2] and retards abscission, which can be used to prevent pre-harvest drop of fruit.[23] Chromosome changes, *e.g.* aneuploidy and polyploidy,[24,25] can occur, but in linseed and peas such abnormalities in the first generation were absent in the second, *i.e.* 2,4-D was modifying but not mutagenic.[26]

2,4-D normally increases the water content of plants,[27] the transpiration loss from leaves being reduced by 25—40%.[2] The herbicide may increase, decrease, or have no effect on mineral-ion uptake and the ion content of the plant.[2,28—30] Overall much depends on the age of the plant and its nutritional status, *e.g.* soy beans grown on soils with high nitrogen content are more sensitive to 2,4-D than those grown on those with low nitrogen content. 2,4-D normally causes a decrease in vitamin content, but 4-chlorophenoxy-acetic acid (4-CPA) has been used to increase the vitamin content of beans by up to 40%.

The colour of red-skinned potato tubers may be heightened by application of 2,4-D[2] but, by contrast, chlorophyll pigments are often greatly reduced, and the flavanoid content of *Convolvulus sepium* is also reduced.[31]

The carbohydrate content of most plants is decreased[32] and even crops of sugar cane show a lower sugar content per acre.[33] Depletion of carbohydrate is caused by a decrease in photosynthesis and an increase in respiration rate.[34,35] Wasteful oxidation of carbohydrate reserves may occur, with an increase in the oxidative activity of the mitochondria.[36,37] Such increased

[16] Z. Kiepal, *Acta Agrobotanica*, 1970, **23**, 73.
[17] B. Sen and J. P. Saini, *Current Sci.*, 1969, **38**, 19.
[18] Ching Chen Wu, T. T. Kozlowski, R. F. Evert, and S. Sasaki, *Canad. J. Bot.*, 1971, **49**, 1737.
[19] J. Hanawa, *Shokubutsugarku Zasshi*, 1968, **81**, 545.
[20] T. A. Mansfield, *New Phytology* 1967, **66**, 325.
[21] G. I. Glushchenko, *Sel'skokhoz Biol.*, 1970, **5**, 37.
[22] S. M. Mashtakov, A. P. Volynets, and A. F. Guz, *Soviet Plant Physiol.*, 1967, **14**, 134.
[23] E. Primo Yufero, *Agroquim. Technol. Aliment.*, 1966, **6**, 360.
[24] B. M. Johri and K. K. Nag, *Current Sci.*, 1968, **37**, 606.
[25] G. H. L. Liang, K. C. Feltner, and O. G. Russ, *Weed Sci.*, 1969, **17**, 8.
[26] M. Aamisepp, *Vaxtodling*, 1966, **22**, 147.
[27] Yu M. Khodorovoski, *Khim. sel'khoz.*, 1967, **5**, 40.
[28] H. M. Etter, *Canad. J. Bot.*, 1967, **45**, 1011.
[29] V. B. Bagaev and S. E. Shkel, *Izvest. Timiryazev Sel'skokhoz Akad.*, 1967, 99.
[30] M. Pannet, *Rev. Fiz. Chim., Ser. A*, 1968, **5**, 425.
[31] R. Tronchet, *Ann. Scient. Univ. Besançon*, 1966, 1.
[32] R. Manteuffel, E. Siegl, and E. Libbert, *Biochem. Physiol. Pflanz.*, 1972, **163**, 586.
[33] J. A. Gumaster and D. B. Barve, Proceedings of the 5th All India Conference of Sugar Cane Research and Development Workers, 1964, 43.
[34] L. W. Rasmussen, *Plant Physiol.*, 1947, **22**, 377.
[35] E. A. Yasnikova, *Nauk Pr. Ukrain. Akad. Sel'skogospod Nauk.*, 1968, **4**, 59.
[36] G. A. Chesalin, V. F. Ladonin, and A. P. Spesivtsez, *Doklady vses. Akad. sel'khoz. Nauk.*, 1968, 9.
[37] S. S. Nair, *Indian J. Exp. Biol.*, 1968, **6**, 182.

oxidative and phosphorylating activities lead to increased amounts of phospholipids and ATP.

After application of 2,4-D the plant may exhibit alteration of the nature of the protein content, there being a great variation in species response.[38–40] The action of 2,4-D on proteins may result in a change in the viscosity of protoplasm, so modifying protoplasmic streaming.[2] Most workers report a large increase in RNA content,[36,41–46] especially messenger and ribosomal RNA, with transfer RNA possibly affected.[47] It appears that the synthesis of RNA is accelerated owing to the enhanced activity of RNA polymerase.[48–50]

Specific reactions of 2,4-D with enzymes may well occur but it seems likely that all the enzyme systems are disturbed. Amongst the enzymes which are known to be stimulated by the herbicide are: oxidative enzymes,[51,52] peroxidase,[42] invertase,[53,54] hydrolase,[54] polyphenol oxidase, RNA polymerase, Indoleacetic acid oxidase, pectin methyl esterase, phosphatase, and glucose 6-phosphate dehydrogenase. By contrast, the following have been inhibited: catalase,[55] ascorbic acid oxidase and peroxidase, cocarboxylase,[51] alanine and ketoglutaric acid transaminase,[56] and carbonic anhydrase.[55] 2,4-D has a multiple effect on the development of indoleacetic acid oxidase isoenzymes, depending on the concentration of the 2,4-D.[57]

Accumulation of toxic metabolites is also caused by 2,4-D.[22] It stimulates the production of ethylene up to 20 to 30 times that of controls,[58–63] and

[38] V. E. Kiselev, *Russ. Physiol.*, 1966, 105.
[39] D. J. Wort and V. S. Rathmore, *Abstracts of 6th Internat. Congress of Plant Proteins*, Vienna, 1967, 439.
[40] V. G. Kholmov, *Agrokhimiya*, 1969, 102.
[41] R. Baxter and J. B. Hanson, *Planta*, 1968, **82**, 246.
[42] S. M. Mashtakov, V. P. Deeva, and V. A. Voilino, *Doklady Akad. Nauk. Belorussk. S.S.R.*, 1968, **12**, 460.
[43] J. A. Zwar and R. Brown, *Nature*, 1968, **220**, 500.
[44] P. G. Jarvis, *Plant Physiol. Lancaster*, 1967, **42**, 52.
[45] R. Baxter, Ph.D. Thesis, University of Illinois, 1967, p. 105.
[46] Y. Masuda and E. Tanimoto, *Plant Cell Physiol.*, 1967, **8**, 459.
[47] M. V. Briquet, R. G. Leblus, and A. L. Wiaux, *Canad. J. Bot.*, 1969, **47**, 1207.
[48] J. B. Hanson and F. W. Slife, Abstracts 155th National Meeting Amer. Chem. Soc., 1968, A9.
[49] T. J. O'Brien, *Diss. Abs.* (B), 1968, **28**, 4846.
[50] T. J. O'Brien, B. C. Jarvis, J. H. Cherry, and J. B. Hanson, *Biochim. Biophys. Acta*, 1968, **169**, 35.
[51] W. K. Kim and R. G. S. Bidwell, *Canad. J. Bot.*, 1967, **45**, 1789.
[52] T. J. Allen and C. L. Leinweber, *Texas Agric. Exp. Station Progr. Report* 1968, 66.
[53] J. Jupy, *Planta*, 1969, **88**, 144.
[54] P. P. Rutherford, E. W. Weston, and A. E. Flood, *Phytochemistry*, 1969, **8**, 1859.
[55] A. Sruginyte and A. Spokiene, *Lieturos T.S.R., Aukst. Mokyklu Mosklo, Darb. Biol.* 1967, **7**, 123.
[56] Z. Chirek, *Ann. Acad. Med., Lodz*, 1967, **9**, 235.
[57] V. V. Chigrin, B. V. F. Koldakov, O. I. Fadeeva, T. P. B. Bortnikova, N. I. Sykalo, and V. I. Zvyagintsev, *Khim. sel'khoz.*, 1970, **8**, 301.
[58] M. Hallaway and D. J. Osborne, *Science*, 1969, **163**, 1967.
[59] R. E. Holm and F. B. Abeles, *Planta*, 1967, **78**, 293.
[60] R. E. Holm and F. B. Abeles, *Plant Physiol., Lancaster*, 1967, **42**, 30.
[61] B. G. Kang and R. P. Ray, *Planta*, 1969, **87**, 206.
[62] F. B. Abeles, *Weed Sci.*, 1968, **16**, 498.
[63] E. C. Maxie and J. C. Crane, *Proc. Amer. Soc. Hort. Sci.*, 1968, **92**, 255.

this may be an important reason for the toxicity of 2,4-D. Tutass[64] proposes a pathway in which oxidative cleavage of a glucose ester of 2,4-D gives monochloroacetic acid, which stimulates $C_2H_4$ production.

In summary, the death of a plant occurs owing to a marked increase in respiration, depletion of energy reserves, lack of photosynthesis, and abnormal growth with aberrant differentiation.

**Animals.**—The feeding of large quantities of 2,4-D to animals results in their death (Table 2), but for man, the only toxic effects have been from the direct ingestion of herbicides for intended suicide or accidental ingestion by children.[65] Of 73 male employees in a factory making 2,4-D, 1.8% had

**Table 2** *Acute oral toxicity of herbicides; single oral doses*[a]

| | $LD_{50}$ *as* mg (kg body weight)$^{-1}$ | |
|---|---|---|
| *Compound* | *Rat*[c] | *Guinea pig* |
| 2,4-D | 375.0—666.0[b] | 1000.0 |
| 2,4-D butyl ester | 1500 | |
| 2,4,5-T | 500.0 | 380.0 |
| silvex | 500.0 | 850.0 |
| piclorain | 8200.0 | 3000.0 |
| 2,3,7,8-tetrachlorodibenzo-*p*-dioxin | 0.022—0.045[b] | 0.0006 |

[a] From proceedings of 23rd Annual California Weed Conference 1971, p. 43.
[b] Varies according to vehicle and sex.
[c] For comparison, the value for 'innocuous' NaCl is 2500 and for 'poisonous' KCN is 1.6.

chloroacne, but no overt clinical case of porphyrinuria was found. Only one worker had persistent uroporphyrinuria, whilst toxicity in other organs was markedly less.[66] Repeated oral doses of 2,4-D 2-ethylhexyl ester at 250 mg kg$^{-1}$ resulted in decreased plasma magnesium and calcium levels in sheep and yearling cattle.[67] Other signs of poisoning were anorexia and reduced weight gain, and often the digestion in the rumen was impaired.[68] Erne[69] reports that renal changes occurred when chickens were given 2,4-D at 1000 p.p.m. in drinking water.

In one study,[70] silvex and 2,4,5-T were given to a ewe at a rate of 25 mg kg$^{-1}$, and after three to four hours a peak concentration of 10 p.p.m. of unmetabolized ester was found. Within 72 hours, 86% of the administered dose was in the urine as unmetabolized ester. With a single oral dose of 25 mg kg$^{-1}$, the maximum concentration of 2,4,5-T in sheep's tissues was 0.1 p.p.m. If, however, repeated oral doses of 250 mg kg$^{-1}$ were given to sheep, levels of up to 368 p.p.m. of 2,4,5-T could be found in the tissues, whereas cattle given

64 H. O. Tutass, *Diss. Abs.* (*B*), 1968, **29**, 877.
65 W. E. Harvey, 23rd Ann. Proc. Calif. Weed Conf., 1971, p. 68.
66 A. P. Polland, *Arch. Environmental Health*, 1971, **22**, 316.
67 L. M. Hunt, B. N. Gilbert, and J. S. Palmer, *Bull. Environ. Contam. Toxicol*, 1970, **5**, 54.
68 J. S. Palmer and R. D. Radeleff, *U.S.D.A. Production Research Reports*, 1969, **106**, 26.
69 N. E. Borklund and K. Erne, *Acta Vet. Scand.*, 1971, **12**, 243.
70 D. E. Clark and J. S. Palmer, *J. Agric. Food. Chem.*, 1971, **19**, 761.

daily oral doses of 0.15 or 0.75 mg $kg^{-1}$ for up to 32 weeks showed no toxic effects, and only low levels of residues were found in the omental fat.

Wright *et al.*[71] examined the conversion of erbon [2-(2,4,5-trichlorophenoxy)ethyl 2,2-dichloropropionate] by sheep into two metabolites 2-(2,4,5-trichlorophenoxy)ethanol (4) and 2,4,5-trichlorophenol (5). High

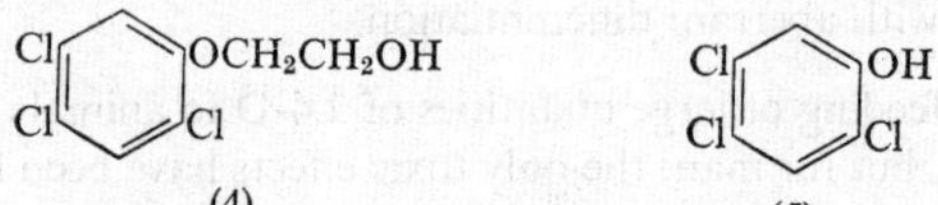


(4) (5)

levels of these metabolites were found in the liver, kidney, and omental fat of sheep given 100 mg $kg^{-1}$, with the highest levels being 6.35 p.p.m. of (4) in the omental fat and 5.54 p.p.m. of (5) found in the kidney.[72] In another study, sheep and cattle fed 2,4-D daily at a rate of 2000 p.p.m. had only 1 p.p.m. or less in the muscle when they were slaughtered on the 28th day.[73] These studies of daily oral ingestion of the herbicide by domestic animals are unlike the situation which occurs in the field. Thus after pastures were sprayed with 2,4-D esters, the milk from cows grazing on them contained 0.01—0.09 p.p.m. of 2,4-D during the first two days after spraying and lower amounts thereafter. When cows were introduced into the pastures 4 days after spraying, residues in milk were below 0.01 p.p.m.[74] In 1969, the Consumer and Marketing Service U.S.D.A. analysed 240 samples of red-meat fatty tissue for 2,4-D, of which 96% showed no residue and only 3 showed more than 0.10 p.p.m. and none as much as 1 p.p.m.[65]

The consensus of opinion seems to be that, at the more usual rates of application, phenoxy-herbicides are not a hazard to animals.[68,75] This opinion is borne out by the example of the accidental aerial spraying of 12 bee colonies. The colonies appeared to be uninjured and no herbicide could be detected in the honey.[76] Also, 2,4-D butyl ester sprayed over forest land in S. Alaska at the rate of 2 lb acid equiv. $(acre)^{-1}$ did not affect the mortality of salmonid fishes or aquatic invertebrates, and the concentrations of the herbicide in water and fish were well below the level considered to be lethal to aquatic organisms.[77]

## 4 Fate of 2,4-D in the Environment

**Plants.**—The action of a herbicide can be subdivided into four ill-defined phases, *viz.* penetration, absorption by the cells, translocation, and finally a toxic action involving the living tissue. The selectivity of 2,4-D in any one plant may depend on interference with any one of these four phases.

[71] F. C. Wright, J. C. Riner, and B. N. Gilbert, *J. Afric. Food. Chem.*, 1969, **17**, 1171.
[72] F. C. Wright, J. C. Riner, J. S. Palmer, and J. C. Schlinke, *J. Agric. Food. Chem.*, 1970, **18**, 845.
[73] D. E. Clark, 162nd National Meeting Amer. Chem. Soc., 1971, Washington, D.C.
[74] D. L. Klingman, C. H. Gordon, G. Yip, and H. P. Burchfield, *Weeds*, 1966, **14**, 164.
[75] E. N. Martinov, *Lesnol Khosyaistvo*, 1970, **6**, 57.
[76] J. O. Moffett and H. L. Morton, *Amer. Bee J.*, 1971, **111**, 382.
[77] H. S. Sears and W. R. Meehan, *Nat. Pesticide Monitor. J.*, 1971, **5**, 213.

In the penetration phase 2,4-D must traverse the lipophilic wax layer, then the cutin, then pectin, and finally the hydrophilic cellulose of the cell walls. The esters of 2,4-D should dissolve in the wax layer more easily and so penetrate the surface faster. Bukovac,[78] Martin,[79] and Hamilton[80] with their respective co-workers have examined the surface constituents of a number of plants, *Plantago major*, *Chenopodium album*, *Taraxacum officinale*, *Rumex obtusifolius*, *Stellaria media*, *Myosotis arvensis*, and *Allium porrum*, with a view to comparing wax composition and 2,4-D absorption. Baker and Bukovac[78] claim that sorption of 2,4-D by the cuticular membrane is inversely related to the amount of cuticular wax and that certain hydrocarbons and aldehydes that are constituents of the wax are more important than others in reducing the penetration of the water in a model system.

Once within the cells, protein binding of 2,4-D[81] may result in the cessation of absorption altogether, which could serve as a means of detoxication. 2,4-D is then translocated in the symplast (the phloem transportation system) and this results in the herbicide being spread throughout the plant, as shown by Figure 1.[82] It has been appreciated for many years that certain features of phenoxyacetic acid explain its functioning as an auxin. At one time a two-point attachment scheme (Scheme 2) was postulated.[83] More recently, van

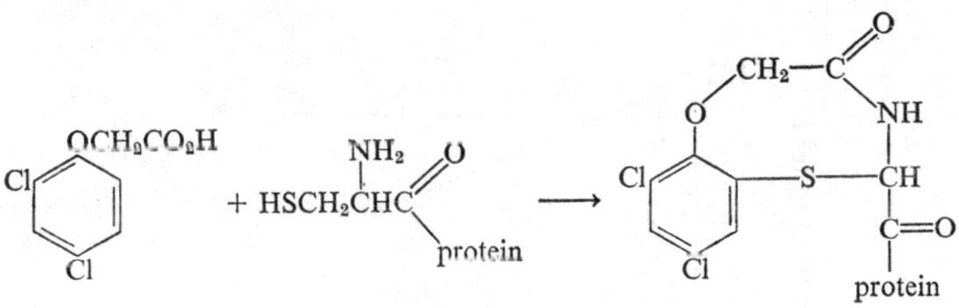


**Scheme 2**

Overbeek[84] suggested that to be effective the auxin molecule should have the following properties:

(*a*) an acid group

(*b*) an aromatic ring or an arrangement of atoms equivalent to a ring, *e.g.* dithiocarbamates.

(*c*) the acid group must be able to assume a position outside the plane of the ring.

(*d*) a balance between the hydrophilic and the hydrophobic parts of the molecule. It is believed that such properties are required because the cell

[78] E. A. Baker and M. J. Bukovac, *Ann. Appl. Biol.*, 1971, **67**, 243.
[79] A. M. S. Fernandes, E. A. Baker, and J. T. Martin, *Ann. Appl. Biol.*, 1964, **53**, 43.
[80] J. E. Allebone and R. J. Hamilton, *J. Sci. Food. Agric.*, 1972, **23**, 777.
[81] V. A. Zemskaya, *Soviet Plant Physiol.*, 1971, **18**, 626.
[82] D. I. Chkanikov, A. M. Makeev, N. N. Pavlova, and V. P. Dubovoi, *Soviet Plant Physiol.*, 1971, **18**, 1067.
[83] R. M. Muir and C. H. Hansch, *Ann. Rev. Plant Physiol.*, 1955, **6**, 157.
[84] J. van Overbeek, *Botan. Rev.*, 1959, **25**, 269.

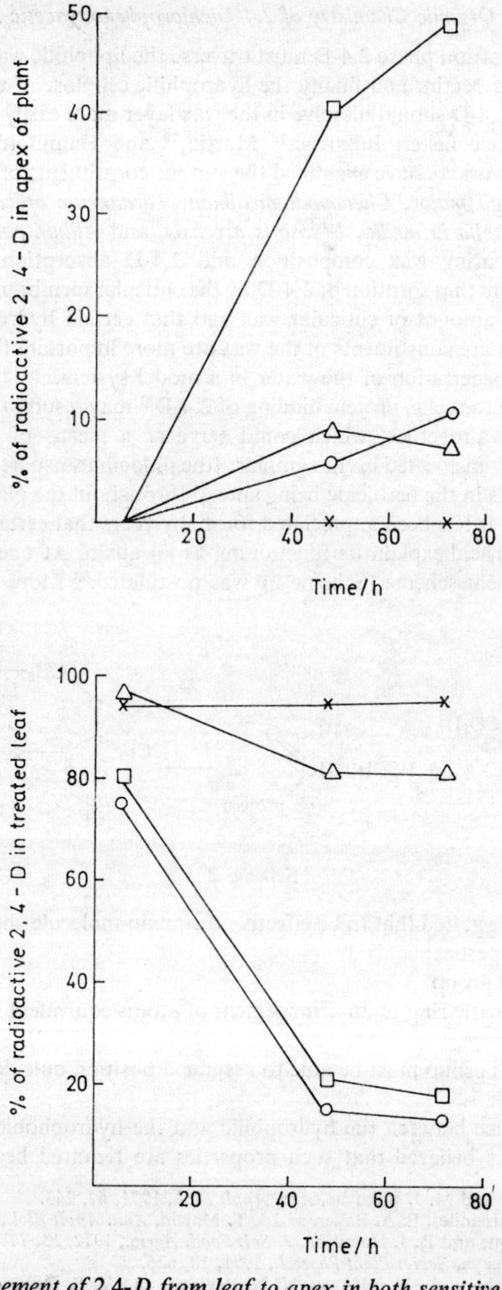


**Figure 1** *Movement of 2,4-D from leaf to apex in both sensitive and resistant plants. Sensitive plants:* ○ *mustard,* □ *sunflower; resistant plants:* × *wheat,* △ *plantain.*

(Based on data in *Soviet Plant Physiol*, 1971, **18**, 1067)

contains slots of definite dimensions into which the auxin molecule must fit, with the active side-chain lying within the hydrogen-bond system of the cell membrane.

The metabolism of 2,4-D within the plant proceeds by three possible mechanisms: either side-chain degradation or ring hydroxylation alone, or ring hydroxylation plus conjugation, or conjugation alone. Although side-chain degradation has been noted in many plants, it appears to be an important detoxication mechanism in only a few, *e.g.* strawberry,[85] garden lilac,[85] and certain *Malus* and *Fragaria* species.[86] The mechanism of side-chain cleavage is not known, though two alternative pathways may operate. The removal of the $C_2$ unit intact is indicated by the presence of 2,4-dichlorophenol in strawberry,[85] bean,[87] sunflower,[87] maize,[87] and barley.[87] The rate of release of $CO_2$ by tick beans was the same whether carboxy-labelled or methylene-labelled 2,4-D was applied, suggesting that the $C_2$ unit is removed intact.[88] Fleeker[89] has now provided evidence for the production of glyoxylic acid (6) by *Ribes sativum*, a resistant species.

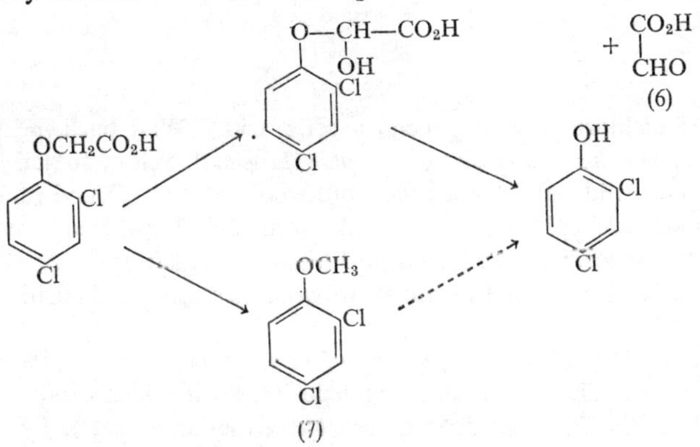

By contrast, most plants release labelled $CO_2$ faster from [1-$^{14}$C]2,4-D than from [2-$^{14}$C]2,4-D. Such stepwise breakdown would suggest an intermediate ether of the type (7) suggested by Luckwill and Jones.[85,86] For many years it has been known that hydroxylated metabolites are present in plants treated with 2,4-D,[90] but it is only recently that success has been achieved in the identification of the metabolites.

4-Hydroxyphenoxyacetic acid is formed when non-chlorinated phenoxyacetic acid is applied to oats, barley, corn, wheat, and pea, but it is not produced by peanuts, soyabeans, and alfalfa.[91] Oats are capable of hydroxylating

[85] L. C. Luckwill and C. P. Lloyd-Jones, *Ann. Appl. Biol.*, 1960, **48**, 626.
[86] L. C. Luckwill and C. P. Lloyd-Jones, *Ann. Appl. Biol.*, 1960, **48**, 613.
[87] D. I. Chkanikov, N. N. Pavlova, and D. F. Gertsuski, *Khim. v. Sel'skom. Khoz.*, 1965, **3**, 56.
[88] M. J. Canny and K. Markus, *Austral. J. Biol. Sci.*, 1960, **13**, 486.
[89] J. R. Fleeker, *Phytochemistry*, 1973, **12**, 757.
[90] R. W. Holley, F. P. Boyle, and D. B. Hand, *Arch. Biochem. Biophys.*, 1950, **27**, 143.
[91] M. Wilcox, D. E. Moreland, and G. C. Klingman, *Physiol. Plantarum*, 1963, **16**, 565.

2-chlorophenoxyacetic acid and 2,6-dichlorophenoxyacetic acid at the 4-position, probably with conjugation, but when chlorine is present at the 4-position, as in 2,4-D and 4-CPA, no hydroxylation occurs.[92] At the other extreme, bean plants hydroxylate 2,4-D with the production of 2,5-dichloro-4-hydroxyphenoxyacetic acid (8) and 2,3-dichloro-4-hydroxyphenoxyacetic acid (9) as the major and minor metabolites, respectively, whilst 2,4,5-T

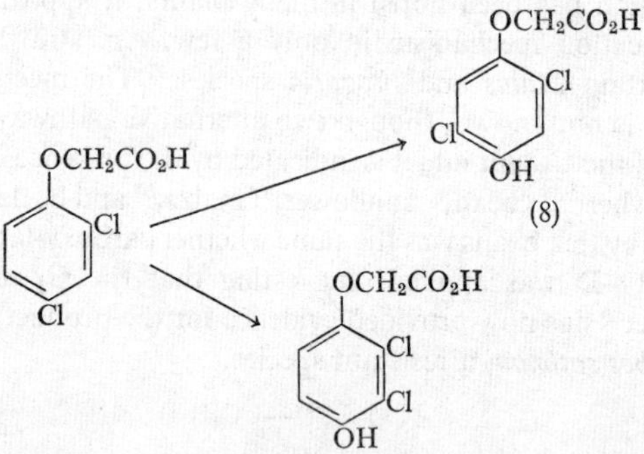


(8) (9)

yields 2,5-dichloro-4-hydroxyphenoxyacetic acid.[93] Wild buckwheat (*Polygonum convolvulus*), leafy spurge (*Euphorbia esula*), yellow foxtail (*Setaria glauca*), and wild oat (*Avena fatua*) produced (8) and (9) and in addition 2-chloro-4-hydroxyphenoxyacetic acid from 2,4-D, whereas very little hydroxylation occurred in wild mustard (*Sinapis arvensis*, syn. *Brassica kaber* var. *pennatifida*), perennial sowthistle (*Sonchus arvensis*), and kochia (*Kochia scoparia*).[94]

Hydroxylation of phenoxyacetic acids may serve as a detoxication method[92,95] but direct evidence is inconclusive. Such evidence as is available is based on the fact that several resistant Gramineae species hydroxylate phenoxyacetic acid whereas some susceptible legumes cannot.[92]

Three types of conjugation have been found in plants treated with 2,4-D, *i.e.* to produce amides (10), sugar esters (11), and glycosides (12), as shown in Scheme 3.

Recently, Feung *et al.*[96] have shown that amides of glutamic acid (10) are formed. Previously only amides of aspartic acid had been reported, in wheat,[97] peas,[98] and red and black currants (*Ribes sativum* and *R. nigrum*).[86] The glutamic acid conjugate was isolated for the first time from soyabean cotyledon callus tissue cultures, although its presence had been postulated in pea epicotyls.[98]

[92] E. W. Thomas, B. C. Loughman, and R. G. Powell, *Nature*, 1964, **204**, 286.
[93] E. W. Thomas, B. C. Loughman, and R. G. Powell, *Nature*, 1964, **204**, 884.
[94] J. Fleeker and R. Steen, *Weed Sci.*, 1971, **19**, 475.
[95] E. W. Thomas, B. C. Loughman, and R. G. Powell, *Nature*, 1963, **199**, 73.
[96] C. Feung, R. H. Hamilton, and F. H. Whitman, *J. Agric. Food. Chem.*, 1971, **19**, 475.
[97] H. D. Klambt, *Planta*, 1961, **57**, 339.
[98] W. A. Andreae and N. E. Good, *Plant Physiol.*, 1957, **32**, 566.

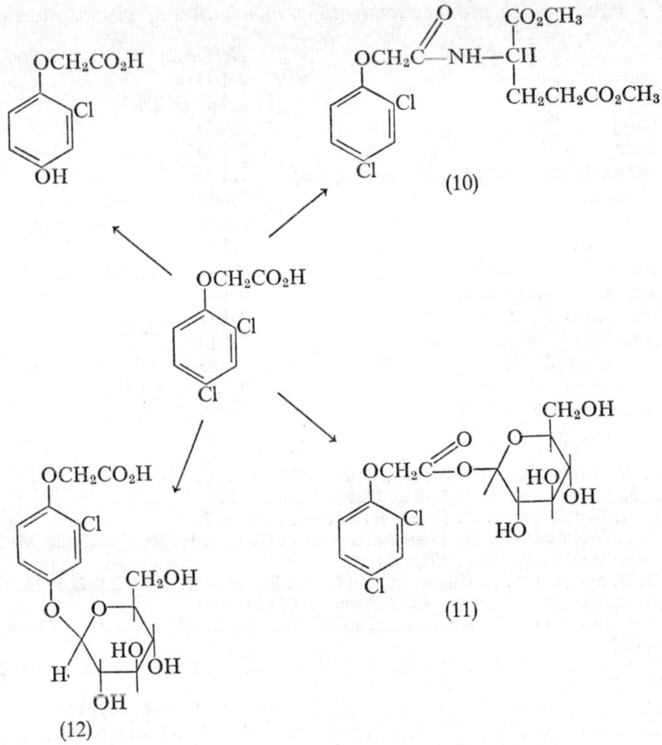

**Scheme 3**

The formation of a glucose ester of 2,4-D, *e.g.* (11), has been reported in wheat[97] and similar esters of 2,4-D, 4-CPA, and 2,6-D in oats.[92] Glycosides of the type (12) are presumed to be the method of conjugation with the phenolic metabolites of 2,4-D.

Freed *et al.*[99] claim that only a small amount of 2,4-D is detoxified by oxidation of the side-chain. Thus in corn and bluegrass almost all the applied 2,4-D is rapidly conjugated, whereas in bean the rate of conjugation is slower. They claim that their findings strongly indicate that inactivation through conjugation is an important factor in selectivity. Conjugation by aspartic acid does not necessarily confer resistance to 2,4-D on plants. Krewson[100] has shown that amino-acid conjugates are active as auxins, which has been confirmed by Feung for soyabean callus tissue. Although there is some evidence that conjugation is an important detoxication route in corn and

99 M. L. Montgomery, Y. L. Chang, and V. H. Freed, *J. Agric. Food. Chem.*, 1971, **19**, 1219.

100 C. F. Krewson, C. H. H. Neufeld, T. F. Drake, T. D. Fontaine, J. W. Mitchell, and W. H. Preston, *Weeds*, 1954, **3**, 28.

**Table 3** *Plants which produce unidentified metabolites of phenoxyacetic acids*

| *Plant* | *Herbicide* | *Ref.* |
|---|---|---|
| cotton | 2,4-D | *a, b* |
| big leaf maple | 2,4-D & 2,4,5-T | *c* |
| sorghum | 2,4-D | *a* |
| bean | 2,4-D | *d* |
| Bermuda buttercup (*Oxalis pes-caprae*) | 2,4-D | *e* |
| citrus peel | 2,4-D | *f* |
| lemon fruit | 2,4-D | *g* |
| cherry trees | 2,4-D | *h* |
| red and black currants | 2,4-D & 2,4,5-T | *i* |
| apples and strawberries | 2,4-D | *j* |
| tick beans | 2,4-D | *k* |
| ironweed | 2,4-D | *l* |
| jimson weed | 2,4-D | *m* |
| cucumbers | 2,4-D & 2,4,5-T | *n* |
| blackjack oak | 2,4-D & 2,4,5-T | *o* |
| bedstraw | MCPA | *p* |

[a] P. W. Morgan and W. C. Hall, *Weeds* 1963, **11**, 130;
[b] J. R. Corbett and C. S. Miller, *Weeds* 1966, **14**, 34;
[c] L. A. Norris and V. H. Freed, *Weed Res.*, 1966, **6**, 212;
[d] R. L. Weintraub, J. N. Yeatman, J. A. Lockhart, J. H. Reinhart, and M. Fields, *Arch. Biochem. Biophys.*, 1952, **40**, 277;
[e] N. G. Marinos, F. H. Chapman, and L. H. May, *Austral. J. Biol. Sci.*, 1964, **17**, 631;
[f] W. R. Meagher, *J. Agric. Food Chem.*, 1966, **14**, 599;
[g] E. A. Erickson, B. L. Brannaman, and C. W. Coggins, *J. Agric. Food Chem.*, 1963, **11**, 437
[h] R. L. Weintraub, J. H. Reinhart, R. A. Scherff, and L. C. Schisler, *Plant Physiol.* 1954, **29**, 303
[i] L. C. Luckwill and C. P. Lloyd Jones, *Ann. Appl. Biol.*, 1960, **48**, 613
[j] L. C. Luckwill and C. P. Lloyd Jones, *Ann. Appl. Biol.*, 1960, **48**, 626;
[k] M. J. Canny and K. Markus, *Austral. J. Biol. Sci.*, 1960, **13**, 486
[l] D. L. Linscott and M. K. McCarty, *Weeds* 1962, **10**, 65;
[m] R. C. Fites, F. N. Slife, and J. B. Hanson, *Weeds*, 1964, **12**, 180;
[n] F. W. Slife, J. L. Key, S. Yamaguchi, and A. S. Crafts, *Weeds*, 1962, **10**, 29;
[o] E. Basler, C. C. King, A. A. Bodiei, and P. W. Santelman, *Proc. Southern Weed Conf.*, 1964, **17**, 351;
[p] E. L. Leafe, *Nature*, 1962, **193**, 485.

wheat,[99,101,102] it is not possible to assign the resistance to one specific conjugate.

A number of metabolites of unknown structure have been noted (Table 3).

A different approach to selective control of weeds was suggested by Wain as a result of his studies with higher homologues of 2,4-D, *e.g.* the phenoxybutyric acids. MCPB and 2,4-DB will kill annual nettle and creeping thistle but not clover, celery, carrots, parsnips, flax, and alfalfa.[103,104] These higher homologues can be used with cereals and forage legumes, and with the tolerant corn or rice grown in the vicinity of cotton and grapes, which are susceptible

[101] S. C. Fang and J. S. Butts, *Plant Physiol.*, 1954, **29**, 56.
[102] S. C. Fang, *Weeds*, 1958, **6**, 179.
[103] R. L. Wain, *Ann. Appl. Biol.*, 1955, **42**, 151.
[104] R. L. Wain, *J. Agric. Food. Chem.*, 1955, **3**, 128.

to phenoxyacetic acid herbicides. This selectivity results from the plants' ability to oxidize the long aliphatic side-chain by a $\beta$-oxidation mechanism. Where the side-chain of a higher homologue of 2,4-D has an even number of carbon atoms, $\beta$-oxidation leads to 2,4-D itself, which then has its normal herbicidal action. Where the side-chain has an odd number of carbon atoms, the oxidation produces a carbonic acid ester as the intermediate, which is hydrolysed to 2,4-dichlorophenol (13), an inactive molecule, as shown in Scheme 4. Wain[105] has made an exhaustive study of 18 homologous series of

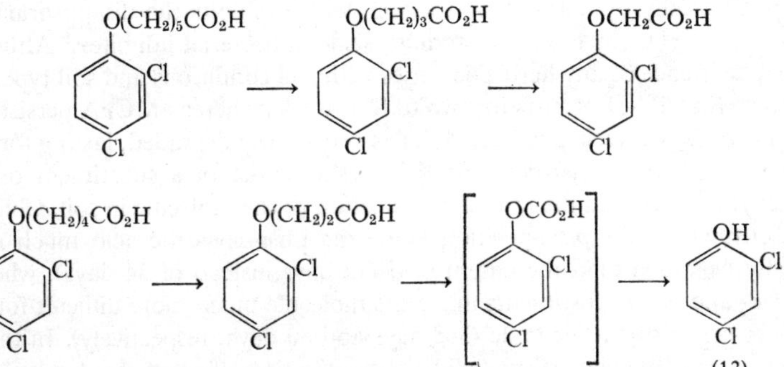


(13)

**Scheme 4**

phenoxyalkanoic acids, whose growth-regulating activity could be related to $\beta$-oxidation of the side-chain.[7]

An unexpected oxidation pathway was also reported by Wain[105] for the degradation of 10-phenoxy-n-decanoic acid in flax. The large amounts of phenol (14) produced could only be explained on the basis of a $\beta$-oxidation

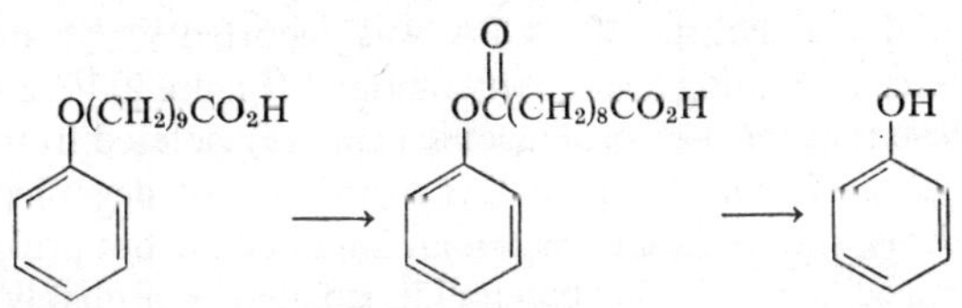


(14)

route. Certain derivatives of 2,4-D have been used as herbicides but it is likely that they act only because they are converted into 2,4-D itself in the plant. Thus isopropyl,[106] butoxyethyl, propyleneglycol butyl,[107] butyl,[108] and octyl[109] esters of 2,4-D have been hydrolysed by corn, castor beans, and cucumber.

[105] C. H. Fawcett, J. M. A. Ingram, and R. L. Wain, *Proc. Roy. Soc.*, 1954, **B142, 60.**
[106] A. S. Crafts, *Weeds*, 1960, **8**, 19.
[107] S. S. Szabo, *Weeds*, 1963, **11**, 292.
[108] C. E. Hagen, C. O. Clagett, and E. A. Helgesen, *Science*, 1949, **110**, 116.
[109] D. J. Morre and B. J. Rogers, *Weeds*, 1960, **8**, 436.

Amides are also hydrolysed to 2,4-D[110,111] whilst long-chain nitriles are either hydrolysed then β-oxidized or β-oxidized and then hydrolysed.[112]

**Soil.**—2,4-D reaches the soil either directly when it is spread as a pre-emergence agent or indirectly *via* the plant. Its fate in the soil is important in that it can affect the soil micro-organisms, which, in turn, influence the soil fertility; it can affect subsequent crops; it can be leached into waterways, where it may be a potential danger to mammals. The degradation of 2,4-D can be shown to be due primarily to the action of soil micro-organisms. This can be shown experimentally by autoclaving the soil, which inhibits the disappearance of 2,4-D,[113—115] or by the use of sodium azide, a bacterial inhibitor.[2] Although the persistence of any herbicide varies with soil conditions and soil type, it is known that 2,4-D persists for two to four weeks whereas MCPA persists for eight to twelve weeks, whilst 2,4,5-T is very slowly degraded, lasting forty to eighty weeks.[2] Audus[2] determined how the nature of a substituent on the phenyl ring affected the rate of detoxication of the molecule in soil. Chlorine substituted in the *para*-position made the phenoxyacetic acid much more easily degraded than the parent acid (12 days instead of 34 days), whereas *ortho*- and *meta*-substituents made the molecule much more difficult for soil micro-organisms to degrade (365 days and 60 days, respectively). In the di- and tri-chlorophenoxyacetic acids, the activating effect of the 4-substituent largely overcame the deactivating effect of other substituents. However, it must be stressed that the same compound behaves very differently in different soils and with different methods of investigation, *e.g.* 2,4-D detoxication can vary from 16 to 94 days in two extreme forms of soil.

A great many soil micro-organisms can degrade the hormone herbicides CPA, 2,4-D, and MCPA. Two main pathways for microbial degradation have been proposed, *viz.* degradation *via* a hydroxyphenoxyacetic acid intermediate or *via* the corresponding phenol.

The bacterial metabolism of 4-CPA was reported[116—118] and has been studied more recently using a soil pseudomonad (Figure 2).[119] Evans and co-workers believe that the 3-chloromuconic acid (15) isolated from the culture may be the *cis-cis*-form. However, this isomer is not degraded further by micro-organisms, showing that it is not an intermediate but probably a more stable rearranged form.[119] The butenolide isolated is probably formed *via*

[110] J. W. Wood and T. D. Fontaine, *J. Org. Chem.*, 1952, **17**, 891.
[111] C. F. Krewson, J. F. Carmichael, T. F. Drake, J. W. Mitchell, and B. C. Swale, *J. Agric. Food. Chem.*, 1960, **8**, 104.
[112] C. H. Fawcett, R. C. Seeley, H. F. Taylor, R. L. Wain, and F. Wightman, *Nature*, 1955, **176**, 1026.
[113] H. R. DeRose and A. S. Newman, *Proc. Soil Sci. Soc. Amer.*, 1948, **12**, 222.
[114] J. W. Brown and J. W. Mitchell, *Bot. Gaz.*, 1948, **109**, 314.
[115] T. P. Hernandez and G. F. Warren, *Proc. Amer. Soc. Hort. Sci.*, 1950, **56**, 287.
[116] W. C. Evans and B. S. W. Smith, *Biochem. J.* 1954, **57**, xxx.
[117] W. C. Evans and P. Moss, *Biochem. J.* 1957, **65**, 8P.
[118] W. C. Evans, J. K. Gaunt, and J. I. Davies, Proceedings 5th Internat. Congress on Biochemistry, Moscow, 1961, p. 306.
[119] W. C. Evans, B. S. W. Smith, P. Moss, and H. N. Fernley, *Biochem. J.*, 1971, **122**, 509.

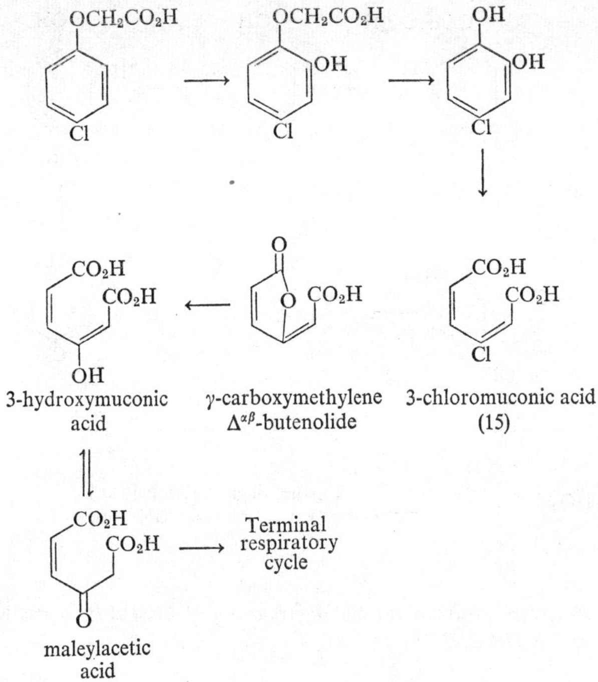


**Figure 2** *Proposed pathway for 4-CPA degradation by a soil pseudomonad*

the *cis-cis*-form.[120] It is believed that the enzyme acting on the culture intermediate is stereospecific towards the *cis-cis*-form of 3-chloromuconate. The degradation of chloro-catechols has been investigated using an *Arthrobacter* sp.,[121–123] and the results helped to confirm previous work.[118]

Studies have been made of the breakdown of MCPA by an *Achromobacter* sp.[124] and *Flavobacterium peregrinum*,[125] both of which liberated 5-chloro-*o*-cresol. Recently, work with a soil pseudomonad N.C.I.B.9340[126] has extended previous studies,[127] leading to the elucidation of a pathway as shown in Figure 3. Although many of the features of this metabolic pathway are similar to that of 4-CPA, the first step in the breakdown of MCPA is the

[120] J. A. Elvidge, R. P. Linstead, and P. Sims, *J. Chem. Soc.*, 1951, 3386.
[121] M. A. Loos, J. M. Bollag, and M. Alexander, *J. Agric. Food. Chem.*, 1967, **15**, 858.
[122] J. M. Bollag, C. S. Helling, and M. Alexander, *J. Agric. Food. Chem.*, 1968, **16**, 826.
[123] J. M. Tiedje, J. M. Duxbury, M. Alexander, and J. E. Dawson, *J. Agric. Food. Chem.*, 1969, **17**, 1021.
[124] T. I. Steenson and M. Walker, *J. Gen. Microbiol.*, 1957, **16**, 146.
[125] J. M. Bollag, C. S. Helling, and M. Alexander, *Appl. Microbiol.*, 1967, **15**, 1393.
[126] J. K. Gaunt and W. C. Evans, *Biochem. J.*, 1971, **122**, 519.
[127] J. K. Gaunt, Ph.D. Thesis, University of Wales, Bangor, 1962.

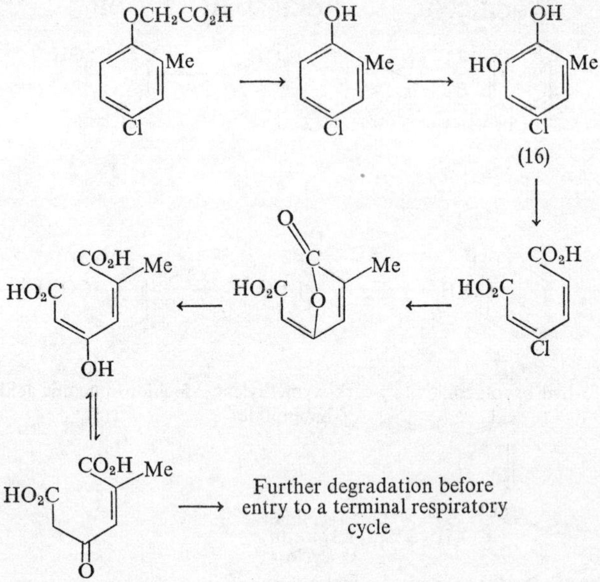


**Figure 3** *Proposed pathway for the degradation of MCPA by a* Pseudomonas *sp., N.C.I.B.*9340

cleavage of the ether link. Such a cleavage of the side-chain could take place by at least three mechanisms:

(*a*) reductive cleavage, giving acetate
(*b*) hydrolytic cleavage, giving glycollate
(*c*) oxidative cleavage, giving glyoxylate

all of which yield the 5-chloro-*o*-cresol (16). Gamar and Gaunt[128] proposed that the pseudomonad species cleaved the side-chain in an oxidative manner. The formation of the lactone could occur by a two-step reaction, *i.e.* dehydrohalogenation followed by lactonization (Figure 4, Route 1 or 2) or *vice versa* (Figure 4, Route 3). Although Route 1 has been observed in bacterial dehalogenation,[129] it is unlikely in the present pathway. Since neither a triply bonded compound (Route 2) nor a halogenated lactone have been detected, Gaunt and Evans[126,130] proposed that a one-step mechanism must operate. Such a reaction is unlike the typical bacterial metabolism of aromatic compounds (lactonization) in that the product contains two double bonds.

Although 2,4-dichloro-6-hydroxyphenoxyacetic acid has been found in cultures of *Pseudomonas* sp. N.C.I.B. 9340 that had been incubated with 2,4-D, no further degradation was noted. The pathway (Figure 5, Route 1)

[128] Y. Gamar and J. K. Gaunt, *Biochem. J.*, 1971, **122**, 527.
[129] J. I. Davies and W. C. Evans, *Biochem. J.*, 1962, **82**, 50P.
[130] J. K. Gaunt and W. C. Evans, *Biochem. J.*, 1971, **122**, 533.

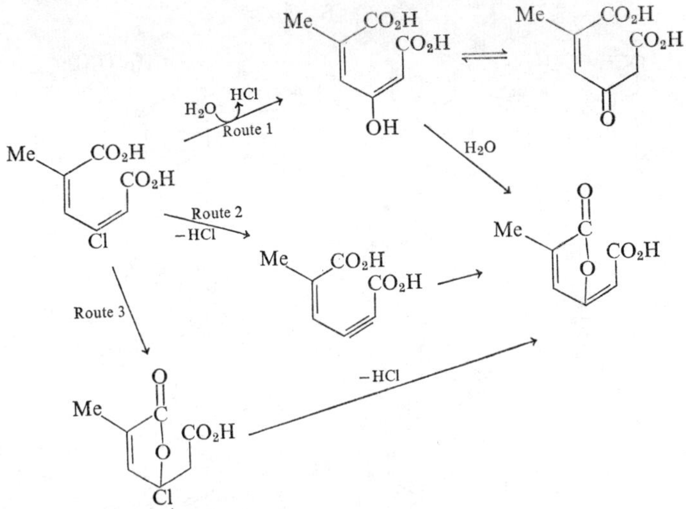


**Figure 4** *Possible two-step lactone formation*

has been suggested as the mechanism for pseudomonads but also for an *Arthrobacter* sp.[123,131] Evans *et al.*[132] and Tiedje *et al.*[123,131] have encountered only the *ortho*-cleavage pathway among microbes performing Route 2 but Horvath[133] has reported the co-metabolism of 4-chlorocatechol and 3,5-dichlorocatechol by a *meta*-cleaving oxygenase from an *Achromobacter* sp. A mechanism similar to that found in higher plants has been suggested by Faulkner and Woodcock[134,135] for degradation by *Aspergillus niger*, in which case 2,5-dichloro-4-hydroxyphenoxyacetic acid and 2,4-dichloro-5-hydroxyphenoxyacetic acid are formed. However, the fungus in incapable of metabolizing these products further.

The degradation of higher homologues of 2,4-D by soil micro-organisms follows the pathways suggested for higher plants, *e.g.* β-oxidation,[136,137] β-oxidation and ring hydroxylation,[138] and cleavage of the ether linkage.[139,140] Breaking of the ring by soil micro-organisms has been followed[141] by studying the u.v. spectrum of the products, though the precise mechanism was not known.

131 J. M. Tiedje and M. Alexander, *J. Agric. Food. Chem.*, 1969, **17**, 1080.
132 W. C. Evans, B. S. W. Smith, H. N. Fernley, and J. I. Davies, *Biochem. J.*, 1971, **122**, 543.
133 R. S. Horvath, *Biochem. J.*, 1970, **119**, 871.
134 J. K. Faulkner and D. Woodcock, *Nature*, 1964, **203**, 865.
135 J. K. Faulkner and D. Woodcock, *J. Chem. Soc.*, 1965, 1187.
136 D. M. Webley, R. B. Duff, and V. C. Farmer, *Nature*, 1957, **179**, 1130.
137 H. F. Taylor and R. L. Wain, *Proc. Roy. Soc.* 1962, **B156**, 172.
138 R. J. W. Byrde and D. Woodcock, *Biochem. J.*, 1957, **65**, 682.
139 I. C. MacRae, M. Alexander, and A. D. Rovira, *J. Gen. Microbiol.*, 1963, **32**, 69.
140 I. C. MacRae and M. Alexander, *J. Bacteriol.*, 1963, **86**, 1231.
141 M. Alexander and M. I. H. Aleem, *J. Agric. Food. Chem.*, 1961, **9**, 44.

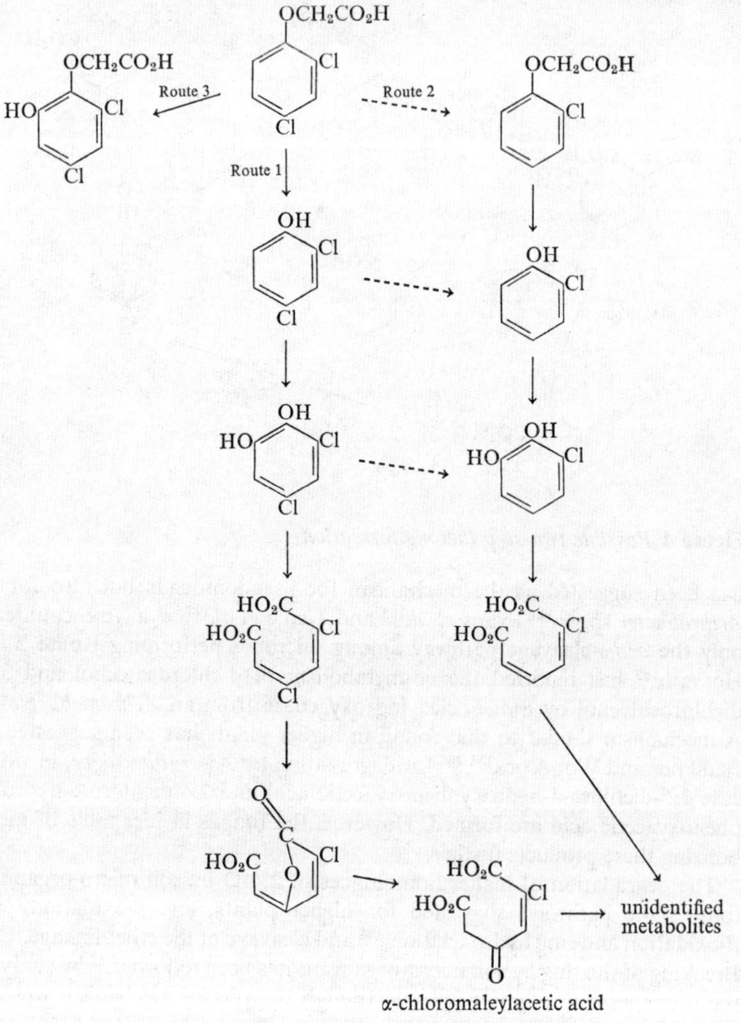


**Figure 5** *Proposed pathway for the degradation of 2,4-D by a soil pseudomonad*

Sodium 2,4-dichlorophenoxyethyl sulphate (sesone) is broken down in the soil by micro-organisms *Bacillus cereus* var. *mycoides*[142–148] to liberate 2,4-D

[142] L. J. King, J. A. Lambrech, and T. P. Finn, *Contrib. Boyce Thompson Inst.*, 1951, **16,** 191.
[143] A. J. Vlitos, *Contrib. Boyce Thompson Inst.*, 1953, **17,** 127.
[144] A. J. Vlitos, *Contrib. Boyce Thompson Inst.*, 1952, **16,** 435.
[145] L. J. Audus, *Nature*, 1952, **170,** 886.
[146] A. J. Vlitos and L. J. King, *Nature*, 1953, **171,** 523.
[147] L. J. Audus, *Nature*, 1953, **171,** 523.
[148] R. B. Carroll, *Contrib. Boyce Thompson Inst.*, 1952, **16,** 409.

itself. Sesone is not effective as a herbicide when sprayed onto the plant foliage.

**Water.**—The most important route for 2,4-D to enter the mammalian system is *via* water. The herbicide itself can reach an aquatic environment either by run-off from soils after agricultural spraying or by spraying lakes and irrigation canals directly to control such plants as the water hyacinth, *Elchornia crassipes*. There is less information about the persistence of 2,4-D in the aquatic environment than on soil persistence, but research in this area is increasing.

In water, herbicides break down either photochemically or under the action of aquatic micro-organisms. Light provides the energy for a variety of degradative reactions including oxidation, reduction, elimination, isomerization, and hydrolysis.[149] The photochemistry of halogenated herbicides[150] has been reviewed. On a laboratory scale, u.v. radiation has been used for the complete removal of organic materials from water samples.[151] Tutass and co-workers[152] have shown that sunlight and u.v. irradiation convert 2,4-D into 2,4-dichlorophenol, 4-catechol, 2-hydroxy-4-chlorophenoxyacetic acid, and 1,2,4-trihydroxybenzene. This latter compound is rapidly oxidized in air and polymerized to a mixture of polyquinoid humic acids. Tutass indicated that such photolytic reaction probably did not occur in plants. Aly and Faust[153] have claimed that photodecomposition accounts for a loss of 50% of a sodium salt of 2,4-D in 50 minutes at pH 7.

Although it is easy to separate the photochemical decomposition of 2,4-D from microbiological degradation in laboratory tests, in natural waters this is not possible, and the action of micro-organisms will be important. Aly and Faust[153] studied 2,4 D breakdown using bottom mud samples from two lakes, one of which had been treated the previous year with 2,4-D, the other being untreated. 2,4-D applied to the mud from the lake treated with it disappeared in 35 days, whereas 2,4-D persisted for 65 days in the other mud, which is in marked contrast to the time needed for degradation in soil of two weeks. In surface waters, levels of 1000 p.p.b. are degraded to 10 p.p.b. within 30 days of application.[154]

De Marco *et al.*,[155] investigating the persistence of 2,4-D in natural lake waters, found that in 'warm' aerobic waters the herbicide was degraded within 6 days, whereas in 'cold' deoxygenated waters it persisted for up to 80 days.

Hemmet and Faust,[156] using domestic activated sewage sludge, decided

[149] D. G. Crosby, Pesticides in the Soil, Int. Symposium, Feb. 1970, Michigan.
[150] J. R. Plimmer, *Residue Rev.*, 1970, **33**, 47.
[151] R. J. Marcus, J. A. Kent, and G. O. Schlenk, *Ind. Photochem. Eng. Chem.*, 1962, **54**, 20.
[152] D. G. Crosby and H. O. Tutass, *J. Agric. Food. Chem.*, 1966, **14**, 596.
[153] O. M. Aly and S. D. Faust, *J. Agric. Food. Chem.*, 1964, **12**, 541.
[154] W. B. House, L. H. Goodson, H. M. Gadberry, and K. W. Dockter, U.S. Dept. of Defence ARPA1086, 1971, p. 369.
[155] J. DeMarco, J. H. Symons, and G. G. Robeck, *J. Amer. Water Works Assoc.*, 1967, **59**, 965.
[156] R. B. Hemmett and S. D. Faust, *Residue Rev.*, 1969, **29**, 191.

that zero-order kinetics applied and that the stoicheiometry was that shown in equation (1).

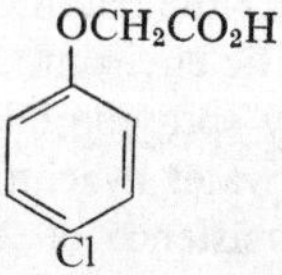

$$+ 7.5O_2 \longrightarrow 8CO_2 + 2H_2O + 2HCl \qquad (1)$$

Negligible quantities of 2,4-D have been found in irrigation water twenty to twenty five miles downstream from the point of soil treatment.[157] The maximum concentration of 2,4-D in water was 25—61 p.p.m. when applied to soil at a rate of 1.9—3 lb $(acre)^{-1}$ and the 2,4-D content was reduced to traces within 30 to 60 minutes. Comes and co-workers[157] are of the opinion that the 2,4-D in the irrigation water could not be considered to be hazardous to animals or crops.

A U.S. Geological Survey of 20 sites[65] on Western streams showed only fractional parts per billion of 2,4-D, 2,4,5-T, and silvex.

Water hyacinth (*Elchornia crassipes*) can be controlled by spraying with the dimethylamine salt of 2,4-D, and it has been claimed that this herbicide is non-cumulative at rates of 22.4—44.8 kg $ha^{-1}$.[158] In a study of water sprayed with 2,4-D amine[159] at 4 lb $(acre)^{-1}$ to control water hyacinth, two sets of water samples had concentrations of 153 and 29 parts per thousand million one hour after spraying, which had changed to 37 and to 30 parts after 24 hours. However, after one week the concentrations had risen to 727 and 1020 parts per thousand million owing to the release of the herbicide from the plants.

Comes[160] claims that dwarf waterlilies, *Nymphaea tetragona*, are better controlled by dichlorobenil alone than in combination with 2,4-D, unlike the control of fragrant water lily (*Nymphaea odorata*).

The levels of 2,4-D in the tissues of fish in treated water are shown in Table 4.[161] Radioactive residues were found in all fish tissues but the 2,4-D content was negligible in muscle, indicating that most of the $^{14}C$ residue was a metabolite. Schultz[161] claims that the major metabolite is a 2,4-D–glucuronic acid conjugate, with the presence of a phenol also noted. Stalling has claimed that there are at least 6 metabolites in fish. Largemouth bass exposed to 0.5, 1.0, or 2.0 p.p.m. of 2,4-D dimethylamine salt in water contained whole-body residues of 1.06, 1.15, and 1.21 p.p.m. after 4, 7, and 14 days exposure.[162] He claims that the effect of herbicides on aquatic ecosystems is short-lived but

[157] P. A. Frank, R. J. Demint, and R. D. Comes, *Weed Sci.*, 1970, **18**, 687.
[158] T. A. Wojtalik, T. F. Hall, and L. O. Hill, *Pesticide Monitor, J.*, 1971, **4**, 184.
[159] K. W. Averitt, Annual Report on Control of Alligator Weed and other Aquatic Plants, 1967, 325.
[160] R. D. Comes and L. A. Morrow, *Weed Sci.*, 1971, **19**, 402.
[161] D. P. Schultz, *J. Agric. Food. Chem.*, 1973, **21**, 186.
[162] C. R. Walker, Proceedings European Weed Research Council, 3rd Internat. Symposium on Aquatic Weeds, 1971, 119.

**Table 4** *Residues* (mg kg$^{-1}$) *of* [$^{14}$C]2,4-*D dimethylamine salt in tissues of channel catfish*[a]

| | *Weeks after start of treatment* | | |
|---|---|---|---|
| *Sample tissue* | 1 | 7 | 12 |
| Blood | 0.10 | 38.74 | 29.95 |
| Brain | 0.17 | 58.63 | 27.25 |
| Bile | 106.49 | 10.97 | 5.04 |
| Liver | 0.77 | 48.31 | 22.17 |
| Striated muscle | 0.06 | 7.33 | 5.63 |
| Lateral-line muscle | 0.05 | 17.08 | 13.47 |

[a] Fish were exposed to a concentration of 2.0 mg l$^{-1}$ of the herbicide for up to 12 weeks.

that the impact of the control of aquatic weeds tends to extend over the whole food chain.

It seems likely therefore that 2,4-D rarely is a hazard in the aquatic environment.

## 5 Analysis

Research on the absorption, translocation, metabolism, and degradation of herbicides is dependent on the accurate measurement of quantities <1 p.p.m. in various tissues and media. One major difficulty in residue analysis is that the interference of any extraneous compounds on the small amount of herbicide present may be tremendous, with the result that efficient and ingenious extraction and purification procedures must be devised. The analysis of 2,4-D may be considered under four headings – extraction, isolation, detection, and quantitative estimation (Figure 6).

**Extraction.**—The extraction of phenoxyacetic acids presents some special problems because they may be present as the free acids, esters, amines, salts, alcohols, or as conjugates with proteins, enzymes, sugars, and other metabolites.[2,163–168] Because 2,4-D and related compounds occur in a variety of media, as shown in Figure 7, there is no standard extraction method. Maceration of the material with a solvent is more efficient than tumbling or Soxhlet extraction.[169] For vegetable matter, a variety of solvents has been used,

163 V. A. Zemshaya and Yu. V. Rakitin, *Soviet Plant Physiol.*, 1967, **14**, 848.
164 D. I. Chkanikov, A. M. Makeev, and N. N. Pavlova, *Khim. Sel'Khoz*, 1968, **6**, 41.
165 Yu. V. Rakitin, V. A. Zemshaya, M. Voronina, and L. M. Chernikova, *Soviet Plant Physiol.*, 1966, **13**, 30.
166 K. Ojima and O. L. Gamborg, *Nippon Dogo-Hiryogaku Zasshi*, 1968, **39**, 484.
167 D. I. Chkanikov and N. N. Pavlova, *Agrokhimiya*, 1966, 115.
168 V. A. Zemskaya, Yu. V. Rakitin, L. M. Chernikova, and Z. V. Kalibernaya, *Agrokhimiya*, 1969, 116.
169 'Analytical Methods for Pesticides, Plant Growth Regulators and Food Additives', ed. G. Zweig, Academic Press, New York, 1963.

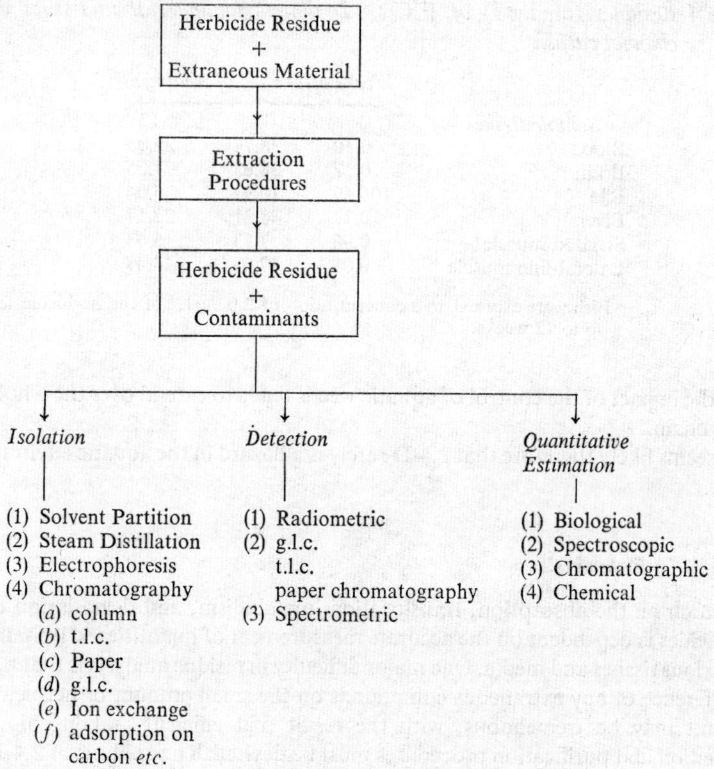


**Figure 6** *Stages of analysis of a sample for the presence of 2,4-D*

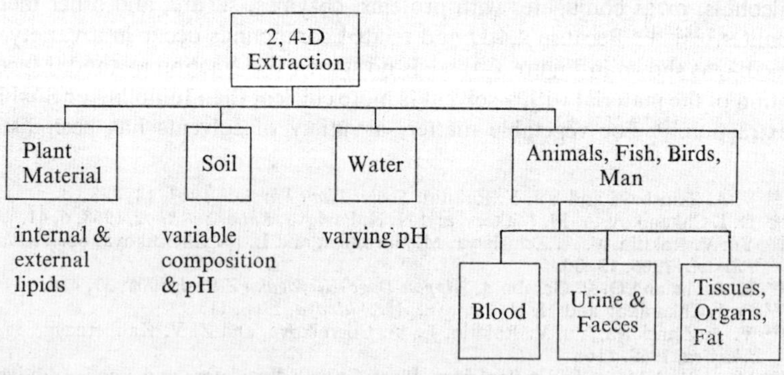


**Figure 7** *Possible media in which 2,4-D may be found*

including ether, acetone, isopropyl alcohol, and water, either alone or acidified.[165,170–173] Mixed systems such as benzene–alcohol,[169] methyl ethyl ketone–ether,[174] isopropyl alcohol–hexane,[175] water–isopropyl alcohol;[176] acidified ethanol–petroleum ether–ether,[169] chloroform–ether–acetic acid,[177] 10% alcoholic sulphuric acid–sodium chloride–water–ether–petroleum ether,[178] 2-propanol–acetone,[179] and water–acetonitrile[180] have also been employed.

Protein-bound phenoxyacetic acids have been hydrolysed enzymically[178] by pepsin for milk samples,[169] and by papain for animal studies.[181] Deproteination using phosphotungstic or phosphomolybdic acids[169,182] can be performed, but simple base hydrolysis is usually sufficient.[183]

A typical plant-extraction technique has been described using chloroform.[183]

Extraction of 2,4-D from soil is less well delineated than for other media,[184] and allowance must be made for binding to soil micelles and the retention of residues by soil flora and fauna. Early extraction methods made no provision for rupturing cells and liberating bound residues.[185] However, an ultrasonic technique has been suggested which will extract more quickly, more reliably, and more quantitatively than conventional methods.[186] Extraction is complete in 30 s compared with the equivalent 8 hour Soxhlet-extraction procedure. Extraction from water follows similar procedures to those used in plant extraction, with careful control of pH.[187–189] Countercurrent extraction has also been used for aqueous media and for animal tissues.[132,190]

**Isolation.**—Fairly extensive purification is necessary following extraction of herbicides from plant or animal matter owing to the presence of large amounts of polar and non-polar, internal and external lipids and pigments. Isolation

170 L. C. Erickson and H. Z. Hield, *J. Agric. Food. Chem.*, 1962, **10**, 204.
171 K. Erne, *Acta Vet. Scand.*, 1966, **7**, 77.
172 V. A. Zemskaya, *Metody Opred. Regul. Rosta. Gerbits Akad. Nauk S.S.R., Inst. Fiziol. Rast.*, 1966, 177.
173 M. I. Gagarina, *Metody Opred. Regul. Rosta. Gerbits Akad. Nauk S.S.R., Inst. Fiziol. Rast.*, 1966, 170.
174 J. Thomson and D. C. Abbot, *R.I.C. Lecture Series*, 1966, No. 3.
175 J. V. Wisnieweski, Hewlett-Packard Applications, Laboratory Report No. 1003.
176 P. L. Pursley and E. D. Schall, *J. Assoc. Offic. Agric. Chemists*, 1965, **40**, 327.
177 R. P. Marguardt and E. N. Luce, *J. Agric. Food. Chem.*, 1955, **3**, 51.
178 G. Yip and R. E. Ney, *Weeds*, 1966, **14**, 167.
179 S. Wathana and F. T. Corbin, *J. Agric. Food. Chem.*, 1972, **20**, 23.
180 M. K. Bach, *Plant Physiol.*, 1961, **36**, 558.
181 D. C. Clark, F. C. Wright, and L. M. Hunt, *J. Agric. Food. Chem.*, 1967, **15**, 171.
182 L. D. Stonov and V. M. Fofanov, *VNII Khim. Sredstv Zaskh. Rast. U.S.S.R., Khim. Sel'khoz*, 1968, **6**, 617.
183 J. E. Allebone, Ph.D. Thesis, Liverpool Polytechnic, 1971.
184 C. E. McKone and R. J. Hance, *J. Chromatog.*, 1972, **69**, 204.
185 H. P. Burchfield, D. E. Johnston, and E. E. Storrs, 'Guide to Analysis of Pesticide Residues', Vol. 1, Office of Public Health, Washington, D.C., 1965.
186 R. E. Johnston and R. I. Starr, *J. Agric. Food. Chem.*, 1972, **20**, 48.
187 T. Kh. Vergeichik, *Vop. Farm*, 1968, 83.
188 A. E. Smith and A. Fitzpatrick, *J. Chromatog.*, 1971, **57**, 303.
189 J. B. Rivers, W. L. Yauger, and H. W. Klemmer, *J. Chromatog.*, 1970, **50**, 334.
190 S. Khanna and S. C. Fang, *J. Agric. Food. Chem.*, 1966, **14**, 500.

is thus best achieved by a combination of solvent-partition and chromatographic procedures.[183,191]

The pH and the solvents required for partition vary depending on whether acidic, neutral, or basic components are being isolated. Clark *et al.*[192] used electrophoresis as well as paper chromatography to study 2,4-D in sheep's urine. Vergeichik[187] found that paper electrophoresis separated 2,4-D from a biological extract better than paper chromatography or t.l.c., and more recently Evans and co-workers identified the $\alpha\gamma$-dichloromuconate metabolite of 2,4-D using this technique.

Column chromatography is frequently used in the isolation procedure since it has the advantage of being able to separate or 'clean up' large quantities of material. Purkayastha[193] used a Florisil column to 'clean up' 2,4-D and 2,4,5-T methyl esters whereas Schultz[161] separated the 2,4-D methyl esters from fish and from water extracts using a silica-gel column and 2% ether in benzene as the eluting solvent. Lambert and Porter[194] built an automatic liquid–liquid chromatographic system for the analysis of insecticides.

For the isolation of herbicides and metabolites from endogenous interferences or from each other, t.l.c. continues to be a major tool. Many procedures in which new substrate (single or mixed), mobile phases, and sensitive, specific chromogenic agents are used have been published.[174] The use of glass, polyester, or foil sheets[195—197] coated in advance with the desired adsorbant substrate greatly facilitates technique and yields more accurate results. The introduction of wedge-shaped plates and channel layer chromatography[198] permits the use of larger samples and eliminates edgewise diffusion of compounds during development or elution of plates. Hamilton and Simpson[199] listed the $R_f$ values of many 2,4-D derivatives for different solvent systems on Florisil thin-layer chromatograms in order to predict their behaviour on a Florisil column.

Paper chromatography gives good results if only small amounts of extraneous material are present. Numerous solvent systems have been used in the identification of phenoxyacetic acids and their metabolites.[200—202]

[191] R. H. Hamilton, J. Hurter, J. K. Hall, and C. D. Ercegovich, *J. Agric. Food. Chem.*, 1971, **19**, 480.
[192] D. E. Clark, J. E. Young, R. L. Younger, L. M. Hunt, and J. K. McLaran, *J. Agric. Food. Chem.*, 1964, **12**, 43.
[193] R. Purkayastha, 163rd National Meeting of Amer. Chem. Soc., Boston, Mass., 1972, PEST 38.
[194] S. M. Lambert and P. E. Porter, *Analyt. Chem.*, 1964, **36**, 99.
[195] E. P. Przybylowicz, W. J. Staudenmayer, E. S. Perry, A. D. Baitsholts, and T. N. Tischer, *J. Chromatog.*, 1965, **20**, 506.
[196] P. Schweda, *Analyt. Chem.*, 1967, **39**, 1019.
[197] H. Thielmann, *Z. analyt. Chem.*, 1972, **262**, 192.
[198] M. J. Matherne and W. H. Bathalter, *J. Assoc. Offic. Agric. Chemists*, 1966, **49**, 1012.
[199] D. J. Hamilton and B. W. Simpson, *J. Chromatog.*, 1969, **39**, 186.
[200] S. L. Vernon and J. Starus, *Phytochemistry*, 1972, **11**, 2723.
[201] K. Randerath, 'Thin Layer Chromatography', Academic Press, New York, 1963.
[202] T. Kh. Vergeichik and E. A. Gryaznova, 'Metody Anal. Pestits', ed. Yu. Lyalikov, Red. Izd-Otd. Akad. Nauk Mold S.S.R., Kishinev, 1970, p. 15.

Ion-exchange resins are not generally recommended for 2,4-D analysis but the anionic resins have been used to absorb acidic herbicides from water extracts of plant tissues. Thornburg[203] found it difficult to obtain quantitative recoveries of materials like 2,4-D from anion-exchange resins, possibly owing to an irreversible reaction of the acidic material with the resin. However, Glennie-Holmes[204] showed that a cation-exchange resin column could be used for the analysis of total acid equivalence of sodium salt pesticides of technical grade. Plant pigments have been removed by adsorbing them on to carbon[173] or Attaclay Nuchar C190N.[171]

Many workers have used radiometric methods for the detection of 2,4-D, especially in the non-quantitative autoradiographs.[205—208] In addition, direct liquid-scintillation radio-assay of $^{14}$C-labelled herbicides in soil has been performed, which proved to be as efficient as a double-extraction process.[209] Phosphorimetry has also been used for 2,4-D analysis.[210]

G.l.c. detection is now primarily by electron capture or microcoulometry, with levels of sensitivity at the picogram level. Other methods have been tried and the levels of their sensitivity are shown (Table 5).[211] For most stationary phases the phenoxy-herbicides and metabolites have to be esterified before g.l.c. analysis. Various techniques for methylation have been recommended, *e.g.* the use of diazomethane,[189,212] diazomethane–methanol–hydrochloric

**Table 5** *Analytical techniques used for phenoxy-herbicides*

| *Method* | *Minimum detectable amount*/g |
|---|---|
| I.r. | $10^{-4}$ to $10^{-12}$ |
| U.v. | $10^{-6}$ |
| G.l.c.–microcoulometric<br>–electron-capture<br>–flame ionization | $10^{-10}$ |
| G.c.–M.s. | $10^{-6}$ |
| M.s. | $10^{-8}$ to $10^{-9}$ |
| T.l.c. prep | $10^{-6}$ |
| G.l.c. prep | $10^{-9}$ |
| paper chromatography | $10^{-6}$ |
| high presure liquid ion-exchange column | $10^{-6}$ |

203 W. W. Thornburg, 'Analytical Methods for Pesticides, Plant Growth Regulators and Food Additives', ed., G. Zweig Academic Press, New York, 1963, p. 108.
204 M. Glennie Holmes, *Pesticide Sci.*, 1972, **3**, 681.
205 S. Wathana, F. T. Corein, and T. W. Waldrop, *Weed Sci.*, 1972, **20**, 120.
206 V. M. Bhan, E. W. Stoller, and F. W. Slife, *Weed Sci.*, 1970, **18**, 733.
207 J. R. Fleeker, *Phytochemistry*, 1973, **12**, 757.
208 H. L. Martin, F. S. Davis, and M. G. Merkle, *Weed Sci.*, 1968, **16**, 88.
209 T. L. Lavy, C. G. Messersmith, and H. W. Knocke, *Weed Sci.*, 1972, **20**, 215.
210 L. B. Sanders and J. D. Winefordner, *J. Agric. Food. Chem.*, 1972, **20**, 166.
211 W. Ebing, *Mitt. Biol. Bundesanst Land-Forstwirt, Berlin–Dahlem*, 1970, **138**, 268.
212 G. Yip, *J. Assoc. Offic. Agric. Chemists*, 1971, **54**, 966.

acid,[213] $BF_3$–MeOH,[169,214,215] dimethyl sulphate,[216] anhydrous methanol with an acidic or basic catalyst,[217] and 3-methyl-*p*-tolyltriazine.[218]

Possibly due to the fact that the methyl esters have relatively short retention times and that they sometimes have similar retention times to co-extractives,[184] several other esters have been prepared, *viz.* chloroethyl,[219] propyl,[220] n-butyl,[184] butoxyethyl,[221] halogenated dimethylsilyl,[222] and trimethylsilyl esters,[222—224] using various esterifying agents. More recently, Woodham *et al.*[219] have reported an improved method for soil analysis incorporating the use of a 10% $BCl_3$ + 2-chloroethanol reagent for esterification, the technique having a sensitivity of 0.01 p.p.m. for 2,4-D.

The use of t.l.c. is well established for the detection of herbicide residues in plant, soil, and water samples[225—227] and for the identification of active ingredients in mixed herbicide formulations.[228] Smith and Fitzpatrick[229] have described a method for the detection of 19 herbicides, including 2,4-D, 2,4-DP, MCPA, MCPP, 2,4-DB, 2,4,5-T, and MCPB, which can be collectively detected in soils at the 1 p.p.m. level and in natural waters at the 0.1 p.p.m. level. More recently, Helling[230] categorized forty pesticides according to their relative mobilities using 'soil' t.l.c. He observed that:

(i) methyl esterification greatly reduced the mobility of the herbicides;
(ii) metabolites of 2,4-D were less mobile than 2,4-D itself;
(iii) the mobility of acidic compounds was directly related to soil pH; and
(iv) two chlorinated *p*-dioxins, potential 2,4,5-T impurities, were immobile.

Geike[231] has shown that 2,4-D, 2,4,5-T, and a number of their analogues could be identified on thin-layer chromatograms at the 2—10 $\mu$g level by their inhibition of acid phosphatase, using nitrophenyl phosphate as substrate.

Many of the t.l.c. methods are very sensitive and approach at least the less sensitive g.l.c. methods for efficiency. T.l.c., paper chromatography, and g.l.c. have the advantage of being able to detect, isolate, and identify a particular

[213] S. F. Howard and G. Yip, *J. Assoc. Offic. Agric. Chemists*, 1971, **54**, 970.
[214] J. R. Baur, R. W. Bovey, and J. D. Smith, *Weed Sci.*, 1969, **17**, 567.
[215] M. G. Merkle and F. S. Davies, *Proc. Soc. Weed Conf.* 1966, **19**, 557.
[216] J. E. Scoggins and E. H. Fitzgerald, *J. Agric. Food. Chem.*, 1969, **17**, 156.
[217] C. J. F. Botcher, F. P. Woodford, E. Bolsma Van Houte, and C. M. Van Gent, *Rec Trav. chim.*, 1959, **78**, 794.
[218] J. L. Landry and C. W. Collier, 157th Meeting, Amer. Chem. Soc., Minneapolis, 1969.
[219] D. W. Woodham, W. G. Mitchell, C. D. Loftis, and C. W. Collier, *J. Agric. Food. Chem.*, 1971, **19**, 186.
[220] G. Yip, *J. Assoc. Offic. Agric. Chemists*, 1971, **54**, 343.
[221] W. R. Meagher, *J. Agric. Food. Chem.*, 1966, **14**, 599.
[222] C. A. Bache, L. E. St. John, and D. J. Lisk, *Analyt. Chem.*, 1968, **40**, 241.
[223] T. P. Garbrecht, *J. Assoc. Offic. Agric. Chemists*, 1970, **53**, 70.
[224] J. R. Baur, R. D. Baker, and F. S. Davis, *J. Assoc. Offic. Agric. Chemists*, 1971, **54**, 713.
[225] D. C. Abbott, H. Egan, W. E. Hammond, and J. Thomson, *Analyst*, 1964, **89**, 480.
[226] D. C. Abbott, K. W. Blake, K. R. Tarrant, and J. Thomson, *J. Chromatog.*, 1967, **30**, 136.
[227] D. C. Abbott, J. A. Bunting, and J. Thomson, *Analyst*, 1965, **90**, 365.
[228] D. C. Abbott and P. J. Wagstaffe, *J. Chromatog.*, 1969, **43**, 361.
[229] A. E. Smith and A. Fitzpatrick, *J. Chromatog.*, 1971, **57**, 303.
[230] C. S. Helling, *Proc. Soil Sci. Soc. Amer.*, 1971, **35**, 732.
[231] F. Geike, *J. Chromatog.*, 1972, **72**, 333.

component in one stage, using retention data and $R_f$ values as well as specific colour reactions *in vitro*. A method of detection devised by Beroza[232] utilizes a combination of solvent partition and g.l.c. to identify a specific compound, based on the fact that no two pesticides are likely to have the same $\rho$ values in a number of solvent systems. These results are shown in Table 6.

**Table 6** *Values of* $\rho$ *for 2,4-D and 2,4,5-T in different solvent systems*

| *Compound* | *Hexane–* MeCN | *Iso-octane–* DMF | *Iso-octane–* 80% *acetone* | *Heptane–* 90% EtOH |
|---|---|---|---|---|
| 2,4-D | 0.02 | 0.01 | 0.16 | 0.09 |
| 2,4,5-T | 0.02 | 0.02 | 0.18 | 0.14 |

U.v. spectroscopy was one of the earliest instrumental methods of analysis of 2,4-D,[233] utilizing the carbonyl absorption band at 220—230 nm and the benzene-ring band at approx. 280 nm. The method has been used qualitatively and quantitatively although it has the disadvantage of being non-specific, and so phenols, aromatic acids, and plant tissues interfere. Muller[234] determined phenoxyacetic acids using u.v. spectroscopy after converting the acids into the methyl esters.

Gore and co-workers[235] studied 2,4-D and other herbicides using u.v. and i.r. techniques, as did Evans and co-workers[132,236] in their metabolic studies. Tomati[237] followed the degradation of 2,4-D by an enzyme complex using u.v. spectrometry. I.r. spectrophotometry can give quantitative as well as qualitative information on both macro- and micro-quantities of phenoxyacetic acids. Malina[238] used i.r. spectroscopy for mixtures of 2,4-D and dicamba whilst Goza[239] has listed i.r. spectroscopic methods for the identification of 60 pesticides, including 2,4-D and 2,4,5-T. Polarographic determination of 2,4-D in biological material has been carried out by Fidelus and Zielek[240] whilst mass spectrometry has been used to identify 2,4-D conjugates as well as the metabolites of 2,4-D, such as succinic acid.[241—243]

**Quantitative Estimation.**—Though bio-assay is not as sensitive as other methods of estimation, it is still an important procedure. One of the earliest bio-assays, relating the growth of corn and cucumber seedlings to varying

232 M. Beroza, M. N. Inscoe, and M. C. Bowman, *Residue Rev.*, 1969, **30**, 1.
233 R. S. Bandurski, *Bot. Gaz.*, 1947, **108**, 446.
234 I. Muller, *J. Agric. Food. Chem.*, 1971, **19**, 1035.
235 R. C. Gore, R. W. Hannah, S. C. Pattacini, and T. J. Porro, *J. Assoc. Offic. Agric. Chemists*, 1971, **54**, 1040.
236 W. C. Evans, B. S. W. Smith, P. Moss, and H. N. Fernley, *Biochem. J.*, 1971, **122**, 509.
237 U. Tomati, D. Lippi, and W. Pietrosanti, *Meded. Fac. Landbouwwetensch, Rijksuniv, Gent.*, 1970, **35**, 829.
238 M. Malina, *J. Assoc. Offic. Agric. Chemists*, 1971, **54**, 706.
239 S. W. Goza, *J. Assoc. Offic. Agric. Chemists*, 1972, **55**, 913.
240 J. Fidelus and M. Zielek, *Mikrochim. Acta*, 1970, 5, 1010.
241 R. D. Hagin, D. L. Linscott, and J. E. Dawson, *J. Agric. Food. Chem.*, 1970, **18**, 848.
242 P. Zink, *Arch. Toxicol.*, 1969, **25**, 1.
243 D. J. Link, *Science*, 1970, **170**, 589.

concentrations of 2,4-D, was developed by Swanson.[244,245] A novel procedure[246] based on the inhibition of enzyme systems by 2,4-D depends on the lipolysis of 4-methylumbelliferone heptanoate (17) to the highly fluorescent alcohol (18). Firm and Wain[247] utilized tomato shoot, cress root, *Avena*

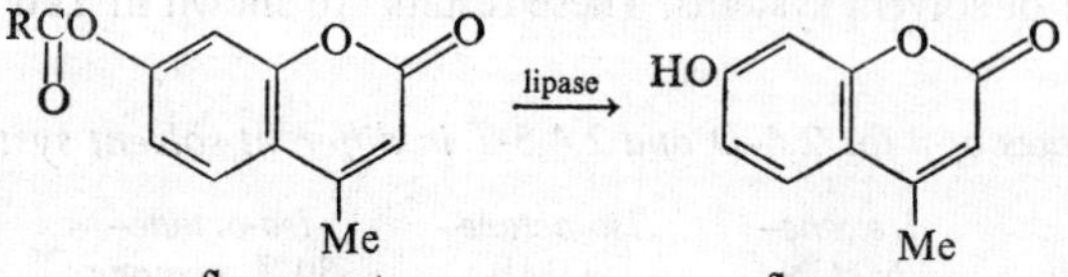

non-fluorescent (17) → fluorescent (18)

mesocotyl, pea segment, wheat cylinder, and *Avena* coleoptile assays during the study of the effect of 3,5-D on plant growth. Kratky and Warren[248] performed three simple, rapid bio-assays (Table 7; *Chlorella* bio-assay, a root bio-assay with sorghum, oat, and cucumber, and a shoot bio-assay with

**Table 7** *Success of three bio-assays for herbicides*

| *Herbicide* | *Concentration/ p.p.m.* | *Shoot*[a] *Sorghum* | *Shoot*[a] *Oat* | *Root*[b] *Sorghum* | *Root*[b] *Oat* | *Root*[b] *Cucumber* | Chlorella[c] |
|---|---|---|---|---|---|---|---|
| 2,4-D | 1 | √ | × | √ | √ | √ | × |
| 2,4-DB | 1 | × | × | √ | × | √ | × |
| 2,4-D | 10 | √ | √ | √ | √ | √ | × |
| 2,4-DB | 10 | √ | × | √ | √ | √ | √ |

√ for inhibition of 50% or greater.
× for inhibition of less than 50%.
[a] Bioassay after 2 days of contact with herbicide; [b] bioassay after 4 days; [c] bioassay after 1 day.

sorghum and oat) on 42 herbicides and they suggest that this is a possible method of primary screening.

Two *in vitro* methods of bio-assays using zoospores have been reported.[249] Other methods of spectroscopic and chromatographic estimation overlap with the work given in the section on detection. Of the chemical methods, chromotropic acid has been used to identify 2,4-D in milk, grain and seed, shellfish, water, and air.

Purely chemical methods are not very sensitive and include fusion with sodium and subsequent determination of NaCl, and Carius oxidation followed by estimation of chloride potentiometrically. A more sensitive technique is cleavage of the ether link and colorimetric estimation of the phenol, obtained by steam distillation, using 4-amino-antipyrine and potassium ferricyanide

[244] C. P. Swanson, *Bot. Gaz.*, 1946, **107**, 507.
[245] C. P. Swanson, *Bot. Faz.*, 1947, **109**, 39.
[246] G. G. Guilbault and M. H. Sadar, *Analyt. Chem.*, 1969, **41**, 366.
[247] R. D. Firm and R. L. Wain, *Ann. Appl. Biol.*, 1971, **68**, 73.
[248] B. A. Kratky and G. F. Warren, *Weed Res.*, 1971, **11**, 257.
[249] R. Hopkins and J. M. Kain, *Marine Pollution Bull.*, 1971, **2**, 75.

Butyl rhodamine and chromotropic acid have also been used, but they are not specific for 2,4-D.

However, g.l.c. is the best method for quantitative estimation of 2,4-D. Most analyses are preformed on silicone stationary phases, *e.g.* OV 17,[161] OV 17 + QF1,[250] and OV1.[123] The sensitivity is now such that 1 nanogram can be determined[250] whilst 1 p.p.b. is the level of detection in water,[158] 0.05 p.p.m. in blood,[251] and 0.5 p.p.m. in muscle.[251] Yip has suggested that instead of devising new methods of analysis we should concentrate on improving those methods which are presently available.

## 6 Conclusion

2,4-D is a popular herbicide; 2,4,5-T, however, is by no means so popular. The oral toxicity of 2,4,5-T is not appreciably different from that of 2,4-D (Table 2). However, on October 29th, 1969, a report on the teratology of 2,4,5-T was published by Bionetics Research Laboratory of Bethesda, Maryland, U.S.A. Several esters of 2,4,5-T caused birth defects in laboratory animals. It was suspected that it was a contaminant of 2,4,5-T which was the culprit, and investigation centred on 2,3,7,8-tetrachlorodibenzo-*p*-dioxin

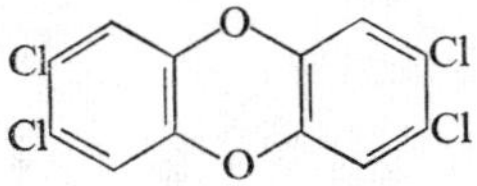


(19)

(19);[05] this substance proved to be very toxic, but tests showed that repeated oral doses of commercial 2,4,5-T did not produce these effects (Table 8).

**Table 8** *Teratology study of phenoxyacetic acids in rats*

| *Treatment* | *No. of litters* | *No. of foetuses* | *Litters with cleft palate* | *Litters with intestinal bleeding* |
|---|---|---|---|---|
| Control (vehicle) | 17 | 194 | 0 | 0 |
| *a* | 17 | 172 | 0 | 0 |
| *b* | 15 | 150 | 0 | 0 |
| *c* | 17 | 173 | 0 | 0 |
| *d* | 17 | 155 | 1 | 8 |
| *e* | 15 | 134 | 0 | 11 |
| *f* | 14 | 76 | 4 | 13 |
| *g* | 15 | 46 | 5 | 8 |

[a] 50 mg pure 2,4,5-T (kg body weight)$^{-1}$ (day)$^{-1}$ in diet;
[b] 0.01 $\mu$g of 2,3,7,9-tetrachlorodibenzo-*p*-dioxin (19) added to 50 mg pure 2,4,5-T (kg body weight)$^{-1}$ (day)$^{-1}$ in diet;
[c] 0.03 $\mu$g of (19) added to 50 mg pure 2,4,5-T (kg body weight)$^{-1}$ (day)$^{-1}$ in diet;
[d] 0.06 $\mu$g of (19) added to 50 mg pure 2,4,5-T (kg body weight)$^{-1}$ (day)$^{-1}$ in diet;
[e] 0.125 $\mu$g of (19) added to 50 mg pure 2,4,5-T (kg body weight)$^{-1}$ (day)$^{-1}$ in diet;
[f] 0.50 $\mu$g of (19) added to 50 mg pure 2,4,5-T (kg body weight)$^{-1}$ (day)$^{-1}$ in diet;
[g] 1.00 $\mu$g of (19) added to 50 mg pure 2,4,5-T (kg body weight)$^{-1}$ (day)$^{-1}$ in diet.

[250] D. W. Woodham, W. G. Mitchell, C. D. Loftis, and C. W. Collier, *J. Agric. Food Chem.*, 1971, **19**, 186.
[251] F. C. Wright, J. C. Riner, and B. N. Gilchrist, *J. Agric. Food Chem.*, 1969, **17**, 1171.

In any event, the Bionetics laboratories' tests had been performed with 2,4,5-T dissolved in dimethyl sulphoxide and administered subcutaneously. However, these tests resulted in the imposition of a ban on 2,4,5-T.[252] The contaminant arises from a side-reaction of normal 2,4,5-T production and is normally present at a level of 1 p.p.m. Thus far it has not been suggested that the contaminant will concentrate in the fat of animals and be subject to biomagnification, as is the case with DDT.

Just outside the coverage of this article, one issue of a new journal has been devoted to 2,3,7,8-tetrachlorodibenzo-*p*-dioxin and its derivatives.[253]

[252] M. Schroeder, *Proc. Western Weed Sci.*, 1972, **25**, 11.
[253] *Environmental Health Perspectives, Experimental Issue 5*, Sept. 1973.

# *Author Index*